Weed Physiology

Volume I

Reproduction and Ecophysiology

Editor

Stephen O. Duke, Ph.D.
Plant Physiologist
Weed Biology and Mechanisms
of Control Research
Southern Weed Science Laboratory
Delta States Research Center
Agricultural Research Service
U.S. Department of Agriculture
Stoneville, Mississippi

CRC Press
Taylor & Francis Group
Boca Raton London New York

CRC Press is an imprint of the
Taylor & Francis Group, an **informa** business

First published 1985 by CRC Press
Taylor & Francis Group
6000 Broken Sound Parkway NW, Suite 300
Boca Raton, FL 33487-2742

Reissued 2018 by CRC Press

Library of Congress Cataloging in Publication Data
Main entry under title:

Weed physiology.

Includes bibliographies and indexes.
Contents: v. 1. Reproduction and ecophysiology—
v. 2. Herbicide physiology.
1. Weeds—Physiology. 2. Weeds—Reproduction.
2. Plants, Effects of herbicides on. 4. Weeds—
Control. I. Duke, Stephen O., 1944—
SB11.W367 1985 632'.58 84-5027
ISBN 0-8493-6313-6 (v. 1)
ISBN 0-8493-6314-4 (v. 2)

A Library of Congress record exists under LC control number: 84005027

Publisher's Note
The publisher has gone to great lengths to ensure the quality of this reprint but points out that some imperfections in the original copies may be apparent.

Disclaimer
The publisher has made every effort to trace copyright holders and welcomes correspondence from those they have been unable to contact.

ISBN 13: 978-1-315-89864-3 (hbk)
ISBN 13: 978-1-351-07774-3 (ebk)

Visit the Taylor & Francis Web site at http://www.taylorandfrancis.com and the
CRC Press Web site at http://www.crcpress.com

PREFACE

Weeds are plants existing at places and/or times at which they are considered undesirable by man. Thus, man's primary interest in weeds is in finding methods for eliminating their presence. Understanding the physiology of weeds and how it differs from that of crop plants is becoming increasingly important in discovering new chemical, genetic, and cultural methods of controlling weeds. The areas of highest interest and most promise in accomplishing this goal are (1) the physiology of weed reproduction, (2) the ecophysiology of weeds, (3) the mechanisms of herbicide action, and (4) the mechanisms of herbicide resistance and tolerance. Volume I deals with the first two of these areas and Volume II discusses the latter two. In these two volumes we have tried to cover each of the above areas at a level that will be useful to weed scientists, plant physiologists, agricultural chemists, and agricultural ecologists. Also, we have tried to provide enough background information for advanced weed science students to effectively utilize these reviews.

In Chapter 1 of Volume I Professor Gutterman discusses flowering, seed development, and influences during seed maturation on seed germination of annual weeds. The emphasis is on how the mother plant and its environment influences seed development and germination characteristics of weed seed. Chapter 2, by Dr. Egley and me, deals with weed seed germination and dormancy. This is an area of tremendous volumes of data, but of little actual biochemical or physiological information at the tissue or cellular level. Where possible, we have tried to discuss the ecological implications of what is known of the physiology. In Chapter 3, by Professors Kigel and Koller, asexual reproduction of weeds is reviewed. This is increasingly becoming an important area of research because annual and perennial weeds with primarily asexual means of reproduction are becoming more troublesome in many agricultural ecosystems. The next two chapters fall into the realm of weed ecophysiology. Dr. Patterson discusses the comparative ecophysiology of weeds and crops and Professor Putnam covers the area of allelopathy as it pertains to weed physiology. Improving our basic understanding of weed physiology in the context of the biotic and abiotic environments within which weeds exist is important in developing a practical knowledge of physiological factors involved in weed competition, weed population dynamics, and other important weed-related processes of agricultural ecosystems.

Volume II deals with the mechanisms of herbicide action and of resistance and tolerance to herbicides. The first five chapters of this volume cover the effects of herbicides and adjuvants on the physiology of plants. Professor Black's chapter begins by covering the effects of herbicides on photosynthesis, including photosynthetic assimilation of nitrogen, sulfur, and phosphorus. This is followed by Dr. Moreland's chapter on herbicide interactions with plant respiration. The third chapter by Professor Bartels deals with the effects of herbicides on chloroplast and cellular development with emphasis on correlating physiological information with ultrastructural effects. My chapter covers herbicidal effects on non-photosynthetic, biosynthetic processes. Nonorganelle membrane functions as affected by herbicides are discussed in Chapter 5 by Professor Balke. In field situations, herbicides are almost always applied with adjuvants. Thus, inclusion of the chapter by Dr. McWhorter on the physiological effects of adjuvants on plants compliments the coverage of herbicides. The last three chapters of the second volume concern the mechanisms of plant resistance and tolerance to herbicides. Professor Gressel's chapter summarizes resistance due to alteration at the site of action of the herbicide. His chapter deals mainly with resistance to herbicides that inhibit photosystem II, since this is the most studied and best understood mechanism of resistance. How absorption and translocation affect tolerance and susceptibility of plants to herbicides is reviewed by Professor Hess (Chapter 8). In the final chapter, Dr. Shimabukuro discusses detoxication of herbicides as a mechanism of tolerance of plants to herbicides.

The rate of increase in the number of publications in the area of weed physiology is growing yearly — a trend seen in many other scientific disciplines. In the case of weed physiology, the increase can be related, at least partially, to the more than 20 billion dollar annual economic loss due to weeds in the U.S. alone and the growing realization that physiological information will be necessary to substantially reduce this loss. For instance, much new research effort is being expended on production of herbicide resistance in crop plants with the techniques of molecular biology. This work can not be intelligently conducted without adequate physiological knowledge of the mechanism of action of the herbicide for which resistance is being sought. Thus, mechanism of herbicide action research is being conducted at a previously unparalleled rate and many of the existing gaps in our knowledge that have been pointed out in these volumes may soon be filled. We hope these reviews will be stimulating and useful to those engaged in this effort and to those interested in plant physiology as it relates to weeds and herbicides.

Stephen O. Duke
February, 1984

THE EDITOR

Stephen O. Duke, Ph.D., is a Plant Physiologist at the USDA's Southern Weed Science Laboratory at Stoneville, Mississippi. He is a member of the Weed Biology and Mechanisms of Control research unit of this Agricultural Research Service laboratory. Dr. Duke received a B.S. in Biology in 1966 from Henderson State College and a M.S. in Botany in 1969 from the University of Arkansas. His education was then interrupted by two years of U.S. Army service, a year of which was spent in Viet Nam. He received a Ph.D. in Botany from Duke University in 1975. That same year he arrived at Stoneville as a National Research Council Associate to study phytochrome and weed seed dormancy. In 1976 he became a permanent staff member of the laboratory.

Dr. Duke is a member of numerous professional societies, including the Weed Science Society of America, the American Society of Plant Physiologists, the American Society for Photobiology, the Japanese Society of Plant Physiologists, the American Chemical Society, and the Scandinavian Society of Plant Physiologists. He is the Vice-Chairman of the Southern Section of the American Society of Plant Physiologists. As a member and chairman of several committees and as an associate editor of Weed Science, he has been an active member of the Weed Science Society of America. In 1984 he was elected the society's outstanding young weed scientist.

Dr. Duke has authored more than 50 research and review papers and has given more than 25 invited lectures at symposia and seminars. His current research interests include herbicide mechanisms of action, plastid development, seed dormancy, plant secondary metabolism, and naturally-occurring compounds as herbicides.

ACKNOWLEDGMENTS

The encouragement and support of Drs. Chester McWhorter and Grant Egley were most appreciated during the editing of these books. I also thank the contributing authors for their time, effort, and patience. Without excellent secretarial assistance, particularly that of Mrs. Judy Fava, my job would have been much more difficult. The encouragement of my colleagues during this task also made the job easier.

The support of my family, particularly that of my wife Barbara, is gratefully acknowledged. Finally, I would like to recognize the many fine teachers that I have had, especially my parents and Professors Adelphia Basford, James Wickliff, and Aubrey Naylor.

CONTRIBUTORS

Nelson E. Balke, Ph.D.
Associate Professor
Department of Agronomy
University of Wisconsin
Madison, Wisconsin

Paul G. Bartels, Ph.D.
Professor
Department of Plant Sciences
University of Arizona
Tucson, Arizona

Clanton C. Black, Jr., Ph.D.
Research Professor of Biochemistry
Department of Biochemistry
University of Georgia
Athens, Georgia

Stephen O. Duke, Ph.D.
Plant Physiologist
Southern Weed Science Laboratory
Agricultural Research Service
U.S. Department of Agriculture
Stoneville, Mississippi

Grant H. Egley, Ph.D.
Plant Physiologist and Research Leader
Southern Weed Science Laboratory
Agricultural Research Service
U.S. Department of Agriculture
Stoneville, Mississippi

Jonathan Gressel, Ph.D.
Associate Professor
Department of Plant Genetics
The Weizmann Institute of Science
Rehovot, Israel

Yitzchak Gutterman, Ph.D.
Associate Professor
Head
Unit for the Introduction and Ecophysiology of Desert Plants
The Jacob Blaustein Institute for Desert Research
Sede Boqer Campus, and Department of Biology
Ben Gurion University of the Negev
Beer Sheva, Israel

F. Dan Hess, Ph.D.
Associate Professor
Department of Botany and Plant Pathology
Purdue University
West Lafayette, Indiana

Jaime Kigel, Ph.D.
Senior Lecturer in Agricultural Botany
Faculty of Agriculture
The Hebrew University of Jerusalem
Rehovot, Israel

Dov Koller, Ph.D.
Professor of Botany
The Hebrew University of Jerusalem
Rehovot, Israel

C. G. McWhorter, Ph.D.
Plant Physiologist and Laboratory Chief
Southern Weed Science Laboratory
Agricultural Research Service
U.S. Department of Agriculture
Stoneville, Mississippi

Donald E. Moreland, Ph.D.
Research Leader
Agricultural Research Service
U.S. Department of Agriculture
Professor of Crop Science, Botany, and Forestry
North Carolina State University
Raleigh, North Carolina

David T. Patterson, Ph.D.
Agricultural Research Service
U.S. Department of Agriculture
Plant Physiologist
Plant Physiology and Photosynthesis Research Unit
Botany Department
Duke University
Durham, North Carolina

Alan R. Putnam, Ph.D.
Professor
Department of Horticulture and Pesticide Research Center
Michigan State University
East Lansing, Michigan

Richard H. Shimabukuro, Ph.D.
Plant Physiologist
Metabolism and Radiation Research Laboratory
U.S. Department of Agriculture
Fargo, North Dakota

TABLE OF CONTENTS

Volume I

Reproduction and Ecophysiology

Volume II

Herbicide Physiology

Chapter 1

FLOWERING, SEED DEVELOPMENT, AND THE INFLUENCES DURING SEED MATURATION ON SEED GERMINATION OF ANNUAL WEEDS

Yitzchak Gutterman

TABLE OF CONTENTS

I. INTRODUCTION

This chapter deals with environmental influences upon flowering mechanisms, seed maturation, and germinability determined during seed maturation of annual weeds. To what extent is the fate of the future generation or generations set by conditions during seed maturation on the mother plant? We have tried to emphasize what is known of the answer to this question, using some of the prevalent types of the most widespread weeds in the world.

The central question examined in this chapter is: What are the ecophysiological adaptations which enable these plants to live in agricultural fields, to persist, and to spread to new areas despite the war waged on them by man? The conditions for survival of the annual weeds are similar in many respects to those of the annual desert plants. Therefore, it is interesting and valuable to compare the annual weeds with the annual desert plants. The flowering mechanisms that enable flowering to occur within a wide range of daylengths, adapting the length of the life cycle according to the daylength, and the maternal and environmental influences on seed development and subsequent germinability are all mechanisms that allow for the distribution of seed germination over time. Even under optimal conditions, only a portion of the seeds germinates at once. This reduces the risk of destruction.

II. FLOWERING

What are the survival mechanisms of some annual weeds that enable them to spread over cultivated fields in very large areas of the world in spite of agricultural efforts to destroy them? Some of the most widespread species will be mentioned here to show their amazing survival potential and their ability to inhabit quite varying habitats in different parts of the world.

A. *Portulaca oleracea* (Portulacaceae)

This is an annual succulent weed found in irrigated fields. It reproduces by seeds and is widely distributed over temperate and tropical regions of the world. It is one of the eight most common weeds on earth.[1]

As reviewed by Kigel and Rubin,[2] the *P. oleracea* species is a polyploid complex. The tetraploid and hexaploid subspecies are usually weeds, whereas the diploid is mainly a halophyte found in coastal areas. *P. oleracea* requires high temperatures and light intensities for its normal development and dies under conditions of shade and frost. Depending on the weather, the complete life cycle can take from 5 weeks to 4 months. In the tropics this plant flowers all year round. The terminal cluster of yellow flowers are usually self-pollinated and develop a globular one-celled dry capsule containing many small black seeds.

P. oleracea under white light was found to be a day neutral (DN) plant in a range of daylengths of 4 to 24 hours (h) of light per day, whereas under extended daylengths with low intensity of white light, this plant was found to be a facultative short-day (SD) plant (Table 1).[2]

In a study by Gutterman,[3,4] the treatment was as follows: 8 h of natural daylight intensity extended to photoperiods of 11, 13, and 15 h with low-light intensity. In a later experiment, the photoperiods were for 16 and 24 h. The plants under 11 h of light develop flower buds in less than 10 days and after the appearance of about three leaves. The longer the daylengths, up to 16 h, the longer the time and the greater the number of leaves until the first flower bud appears (Table 2).[3,4] These experiments had 27°C day and 22°C night temperatures. At lower temperatures (16/11°C) the life cycle is slightly longer.[3]

A short life cycle is a very important survival mechanism of a weed in cultivated areas. An additional mechanism is found in *P. oleracea*. The plant is able to continue seed maturation and produce mature seeds even when it is cut after flowering.

Table 1
ENVIRONMENTAL CONTROL OF FLOWERING OF *PORTULACA OLERACEA*

Locality	Light treatments	Photoperiod response	Ref.
Michigan	White light of 4—24 h	DN	62
Veranesi, India	White light of 4—24 h	DN	1
Israel	Extended daylength with low-intensity white light	FSD	3, 4
India	Same as above	FSD	63

Note: DN = day neutral; FSD = facultative short day.

Adapted from Kigel, J. and Rubin, B., *Handbook on Flowering*, Vol. 4, Halevy, A., Ed., CRC Press, Boca Raton, Fla., 1985. With permission.

Table 2
THE INFLUENCE OF THE DAYLENGTHS ON THE NUMBER OF LEAVES UNTIL FLOWER BUD APPEARANCE AND THE AGE OF THE PLANT AT THE FIRST FLOWER BUD APPEARANCE IN *PORTULACA OLERACEA* PLANTS[a]

Photoperiod treatment (hours of light per day)	Number of leaves until flower bud appearance	Age at first flower bud appearance (days)
8[b]	4.4 ± 0.5	12.0 ± 0.7
11[b]	3.2 ± 0.4	9.7 ± 0.2
13[b]	7.0 ± 0	14.0 ± 0
15[b]	10.5 ± 0.2	19.0 ± 0
16[c]	11.1 ± 0.3	19.0
24[c]	11.1 ± 0.2	18.6

[a] Adapted from Gutterman.[3,4]
[b] Study started on 16-8-66.
[c] Study started on 28-11-67.

A night break of red light (see Figure 1, treatment G) caused a delay of flower bud appearance. Flower buds appeared after 13.8 leaves, a delay in comparison to long-day (LD) conditions (11.3 leaves, treatment B). However, a far-red night break accelerates the appearance of the flower buds. They appear after six leaves (treatment H) in comparison to SD of 8 h of daylight in which the flower buds appear after 7.9 leaves (treatment A). Eight hours of red or far-red light following the 8 h of a dark period caused a delay in flower bud appearance, at 15.8 (treatment I) and 12.1 (treatment J) leaves, respectively. All of the other treatments have nearly the same effect as the SD.[4]

B. *Amaranthus* (Amaranthaceae)

As reviewed by Kigel,[5] *Amaranthus* is a genus of tropical origin, widely distributed all over the world, having about 75 species, most of which are pioneer annuals of disturbed soils. Some of these species are known as weeds. *A. retroflexus* is a facultative SD plant for flowering with a critical daylength of 12 to 14 h. Plants are also able to flower in continuous light. A daily night break of 1 h of fluorescent light inhibits flowering. However, daily irradiation of 20 min of far-red light preceding 16 h of dark periods promotes flowering.

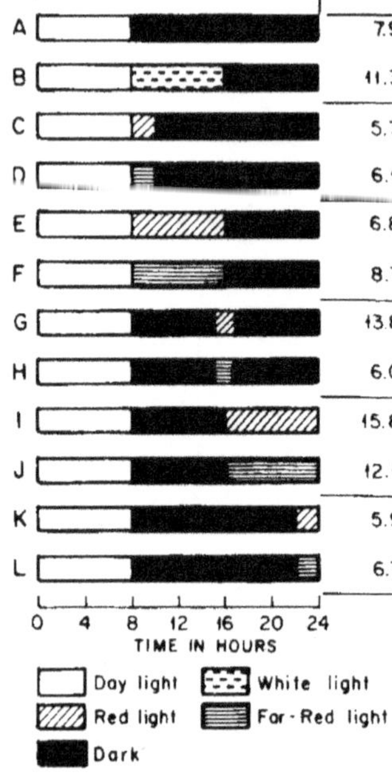

Photoperiodic & spectral conditions of mother plant	Number of leaves of mother plant	GERMINATION PERCENTAGE				
		WHITE LIGHT			RED LIGHT	
		24 hrs	48 hrs	72 hrs	44 hrs	68 hrs
A	7.9 ± 0.2	30.0 ± 1.6*	85.0 ± 0.7	93.0 ± 1.2	87.0 ± 1.8	95.5 ± 1.1
B	11.3 ± 0.2	18.5 ± 1.7	76.5 ± 1.3	80.5 ± 1.4	61.0 ± 3.1	75.5 ± 1.5
C	5.7 ± 0.2	56.5 ± 2.8	98.0 ± 0.5	98.5 ± 0.3	76.0 ± 2.5	95.0 ± 0.6
D	6.1 ± 0.3	14.0 ± 1.2	55.5 ± 1.9	62.0 ± 2.2	31.0 ± 2.6	44.5 ± 1.5
E	6.8 ± 1.0	25.5 ± 2.8	94.5 ± 1.2	98.0 ± 0.5	40.0 ± 1.1	88.0 ± 1.2
F	8.7 ± 0.5	28.0 ± 3.9	90.5 ± 2.0	95.0 ± 1.4	78.5 ± 0.3	91.5 ± 0.9
G	13.8 ± 0.3	29.5 ± 7.0	91.5 ± 1.5	93.0 ± 1.8	76.0 ± 2.9	93.5 ± 1.3
H	6.0 ± 0.2	25.0 ± 1.4	77.0 ± 1.8	86.0 ± 1.7	78.5 ± 2.0	88.5 ± 1.0
I	15.8 ± 0.6	29.0 ± 3.0	92.0 ± 1.1	97.0 ± 1.1	78.0 ± 2.4	92.5 ± 1.2
J	12.1 ± 0.7	34.0 ± 2.6	76.5 ± 2.2	85.0 ± 2.1	56.5 ± 3.6	78.5 ± 2.6
K	5.9 ± 0.3	13.0 ± 1.4	96.0 ± 1.4	98.5 ± 0.6	85.5 ± 1.5	96.0 ± 0.8
L	6.7 ± 0.2	10.5 ± 0.7	83.5 ± 1.8	88.5 ± 1.8	77.0 ± 0.7	95.0 ± 0.7

FIGURE 1. Average germination percentage of seeds of *P. oleracea* (4 batches of 50 seeds each per treatment) germinated under continuous white and red light. The seeds were derived from mother plants grown under various photoperiodic and red/far-red light conditions and were also influenced by the average number of leaves at the time of flower bud appearance. *Standard error. (From Gutterman, Y., *Oecologia*, 17, 27, 1974. With permission.)

Seedlings grown under LD conditions, receiving one SD of 8 h, 3 days after germination, flowered before seedlings in continuous LD of 16 h. This one SD cycle is sufficient to promote flowering initiation. Photoperiods of 8 h of blue light delayed flowering. However, flowering was strongly promoted when blue light was combined with far-red light.[5]

It is interesting to note that the far-red promotion could be a very important survival mechanism for weeds growing under crop canopies where shadow is rich in far-red light (see Section IV.D).

A. retroflexus seedlings are able to flower in a relatively short time under low-light intensities in both SD and LD. Flowering even appears in light that is only 12% of natural sunlight intensity under SD conditions after about 28 days. In 60% light intensity, flowering takes place in 15 days under SD conditions and in 41 days under LD conditions. If light intensity is 27% of the natural light, flowering appears after approximately 61 days under LD (Table 3).[6]

A. hybridus is less sensitive to daylengths than other species in this genus.[5] This plant is a facultative SD plant. Under greenhouse conditions, flowering appears faster under SD and after less leaves are produced than under LD in the temperatures 28/22°C and 22/16°C. These plants flowered in Iowa, U.S. under natural daylengths of 15 to 16 h, and in Israel they flowered under natural daylengths of 13 to 14 h.

In the two weeds *A. retroflexus* and *A. hybridus*, plants that were developed under natural conditions from seeds collected in northern latitudes flower earlier than those from seeds collected in southern latitudes. This is an important survival mechanism to ensure that plants will complete their life cycle and produce at least some seeds before the fall frost. The same was reported about the perennial weed *Sorghum halepense* Pers. (johnsongrass). As reviewed

Table 3
SOLAR LIGHT INTENSITY AND FLOWERING OF *AMARANTHUS RETROFLEXUS* AT 27/22°C IN 16-H AND 8-H PHOTOPERIODS

	Days from sowing to flowering	
Solar light intensity (%)	SD (8 h)	LD (16 h)[a]
60[b]	15.1 ± 0.1	40.7 ± 2.0
43	16.8 ± 0.5	53.0 ± 2.2
27	19.7 ± 0.5	60.7 ± 2.2
12	27.8 ± 0.6	[c]
Final no. leaves on main stem	6.8 ± 0.3	22.8 ± 0.9

[a] Illumination for 16 h daylength extension = 10^4 erg cm^{-2} sec^{-1}.
[b] At light intensity of 60% = 3.9×10^5 erg cm^{-2} sec^{-1} at midday.
[c] After 97 days, none flowered.

Adapted from Koller, D., Kigel, K., Nir, I., and Ofir, M., Environmental Control of Weed Physiology, Final Report U.S.D.A. Project A10-CR-27, U.S. Department of Agriculture, Washington, D.C., 1975.

by Kigel,[7] flowering behavior of *S. halepense* differs in plants from different localities, from facultative SD to day neutral (DN). *Xanthium strumarium* from latitudes between 20° to 45.5°N have a critical nightlength from about 7.5 h in the north to about 10 h in the south.[8] *X. strumarium* plants from the same latitude but different altitudes differ in their photoperiodic responses according to the climatic conditions.[9]

C. *Chenopodium album* (Chenopodiaceae)

C. album flowers under any daylength.[1,10] In Chandigarh, India, Ramakrishnan and Kapoor[11] found and studied two cytological races of *C. album*, a diploid (2n = 18), growing during the summer from April to September. The hexaploid (2n = 54) flowers during the winter from November to March.

Flowers of the summer populations begin to develop during days of 13.5 to 14.0 h and temperatures of 29/31°C. Each seasonal population mentioned above is well adapted, having differences in their germinability and conditions for development and flowering. In the summer population, low temperatures inhibit flowering. In the winter population, high temperatures inhibit flowering. Seasonal ecotypes were found also in *X. strumarium*.[12]

D. Conclusions

Of the ten most common and destructive annual weed species across the world, according to Holm et al.,[1] five out of six (*Echinochloa crusgalli, Eleusine indica, P. oleracea,*[4] *C. album,*[10] and *Rottboellia exaltata*) are mentioned to be DN for flowering while one, (*Digitaria sanguinalis*), is SD for flowering.[1] Information on the flowering response of the other four is not mentioned. *Cyperus rotundus* (purple nutsedge) is found to be intermediate in flowering response to daylength.[13]

The ability of certain weed species to flower over a wide range of daylengths is a very important survival mechanism for weeds throughout the world. Two plants found to have the flexibility to develop geographical ecotypes and genotypes inhabiting the same area in different seasons are *Chenopodium album*[11] and *X. strumarium*.[12] The exceptional ability of adaptation to new habitats enables such weeds to flourish in a large variety of habitats. Some of the desert annual plants that were studied in the Negev Highlands and survived under unpredictable amounts and distributions of precipitation were found to have the ability to flower in a wide range of daylengths. The majority are facultative SD plants and, in some cases, species were found to be DN plants for flowering.[14]

These mechanisms ensure that plants which germinate after a late rain will also be able to produce seeds the same as plants which germinate after an early rain when daylengths are much shorter. The longer the daylength, the shorter the life cycle of the facultative LD plants. This is a safety mechanism which enables the desert winter annuals to finish the life cycle and to produce seeds before the hot, dry summer.[14-17] The facultative behavior of some weeds for flowering ensures that they will germinate late and that they will complete their life cycle, producing seeds before the winter frost.[3-5] A short life cycle enables the weeds to produce seeds between the periods when weed killers are applied.

III. SEED DEVELOPMENT AND MATURATION

During seed maturation in angiosperms, there is an accumulation of storage materials such as starches and proteins in the endosperm and in the embryo. Large changes also occur in other substances such as hormones, enzymes, DNA, RNA, mRNA, etc. This was widely reviewed in Bewley and Black[18] and others.[19-24] The time scale of the cardinal events of seed maturation from anthesis of *Pisum* was reviewed and summarized by Pate (Figure 2).[25] Cell division of the seed embryo cotyledons started before the pod reached the final size. Near the end of the cell division stage, the endosperm in the developing seeds was depleted and there was an increase in the number of cells in the stage of seed enlargement. There was a lag in the stages of development between maximum cell division and maximum cell enlargement when the embryo fills the embryo sac. The cell enlargement stage reached its maximum when the pod fresh weight also reached a maximum. During the stage of cell enlargement, there is an increase in dry matter accumulation of RNA, proteins, and starch. As the pods dry out, there is a decrease in seed fresh weight, RNA, proteins, and starch accumulation.

A correlation was found between the growth of the pods and the hormonal content in the seeds during early seed development, as was seen in *P. sativum* c.v. Alaska (Figure 3).[26] This suggests that the movement of dry matter such as carbohydrates and nitrogenous and other substances into the seed might be regulated by the hormones accumulated in the seed itself.

Changes in the hormonal content, in the liquid endosperm, as well as in the embryo take place in close correlation to seed growth.[27] As one can see from Figure 4, the peak accumulation of auxin and gibberellin in the liquid endosperm occurs 4 to 8 days after full bloom, whereas the abscissic acid (ABA) is at very low levels during this time. A second peak of gibberellic acid (GA) content occurs in the 12th day. The amount of GA in the embryo reaches its peak 16 days after the full blooming. Then a gradual decrease in GA begins. The ABA in the embryo starts to increase rapidly from the 16th day, reaching a peak on the 24th day after the full blooming. The auxin that appears in high amounts in the liquid endosperm, approximately 4 days after full blooming, is found only in very small amounts in the embryo throughout its development (Figure 4).[26] Not only is there a change in the relative amounts of different hormones in the developing seed, but varying quantities of different gibberellins are found during the development of *Phaseolus multiflourus* seeds.[28] In studies of *Pisum sativum* seeds it was shown that "bound" GA, previously stored during seed maturation, seems to be liberated and utilized during germination.[29]

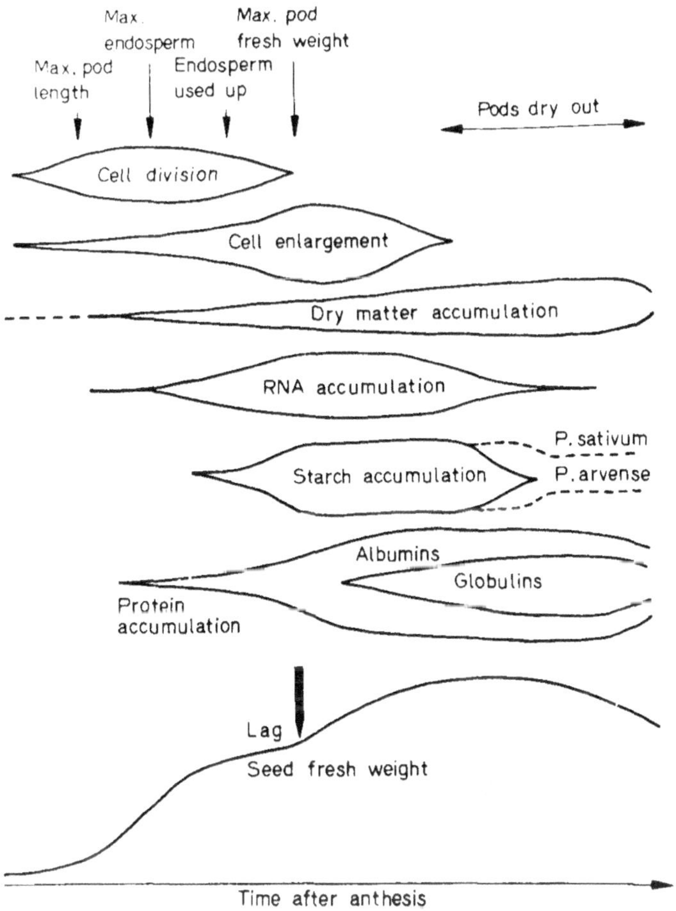

FIGURE 2. Schematic representation of the cardinal events in seed maturation of *Pisum* spp. The common time scale is the growth curve for seed fresh weight. All events depicted refer to rates within the developing cotyledons, divergent lines denoting periods of increasing rate and convergent lines times of decreasing rate. Differences between *P. sativum* and *P. arvense* are indicated. The solid arrow marks the approximate time when seeds of *P. sativum* are harvested for quick freezing. (From Pate, J. S., *Crop Physiology*, Evans, L. T., Ed., Cambridge University Press, New York, 1975, 191. With permission.)

When hormones were applied to the inflorescence of *Lactuca scariola* at the time of flower bud appearance, an increase of one growth substance in the parent plant resulted in the accumulation of a different growth substance in the seeds. Differences were also seen between seeds that had matured under LD and those that had matured under SD (Figure 5; also see Section IV.B).[30] A reduction of the gibberellin content by β-chloro-ethyltrimethyl ammonium chloride (CCC) in *Pharbitis* seeds and its after-effect on the progeny were found by Zeevaart.[31]

Felippe and Dale[32] studied the effect of CCC and GA treatment on *Phaseolus* plants. It was found that an increase in the ABA during seed maturation could have an important role in the prevention of vivipary of the embryo. A reduction of the ABA content occurs at the same time as the decrease in the water content of the embryo (see review by King[21]).

During the process of seed development and maturation, the plant develops an independent organ containing an embryo, storage materials, protective tissue, and sometimes transport

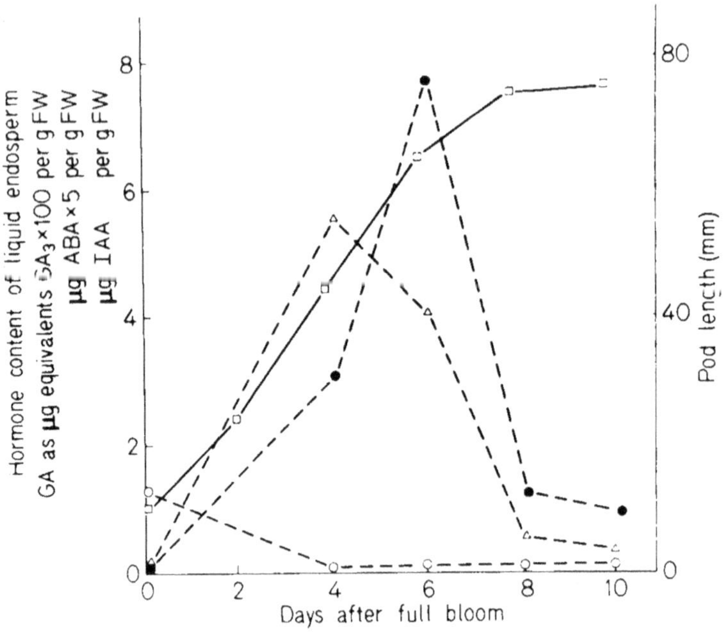

FIGURE 3. Ovary wall growth and hormone content of seeds during early seed development in *Pisum sativum* cv. Alaska. □—□: pod length ○—○: ABA; ●—●: auxin; △—△: gibberellin. (From Eeuwens, C. G. and Schwabe, W. W., *J. Exp. Bot.*, 26, 1, 1975. With permission.)

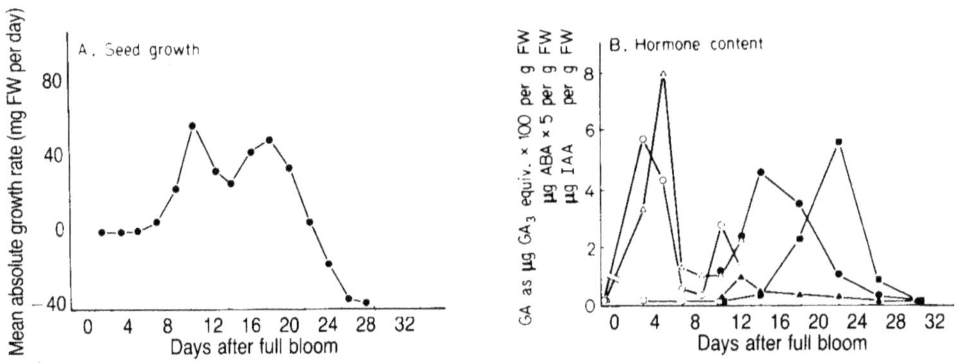

FIGURE 4. Seed development (A) and hormone levels (B) in *Pisum sativum* cv. Alaska. □—□: ABA in liquid endosperm; ○—○: GA in liquid endosperm; △—△: auxin in liquid endosperm; ■—■: ABA in embryo; ●—●: GA in embryo; ▲—▲: auxin in embryo. (From Eeuwens, C. G. and Schwabe, W. W., *J. Exp. Bot.*, 26, 1, 1975. With permission.)

tissues. Seed formation is the only way in which annual plants can ensure continuity in future generations. Most weed seeds have the ability to germinate at the right time and in the right place after remaining for years in the soil seed bank. These seeds also can withstand extreme environmental conditions such as frost or hot and dry seasons without losing their viability. Then, when conditions are right, they will germinate and start a new generation.

During plant development and seed maturation, environmental and maternal factors can affect and change seed germinability and viability. This is of great importance for plant survival.

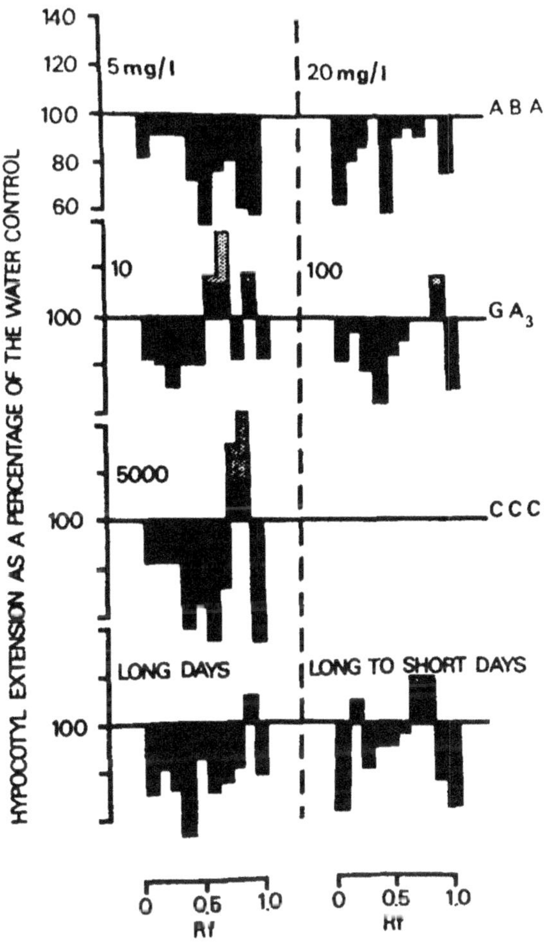

FIGURE 5. Lettuce hypocotyl assays of extracts of seeds harvested from *Lactuca scariola* plants given the treatments shown. All plants treated with chemicals were grown continuously in long days. On the vertical axis, the extension of the water-control epicotyls is given as 100% and extension at each Rf is related to this value. Shaded areas of the histogram indicate promotion significant at $p<0.05$. (From Gutterman, Y., Thomas, T. H., and Heydecker, W., *Physiol. Plant.*, 34, 30, 1975. With permission.)

IV. ENVIRONMENTAL AND MATERNAL INFLUENCES DURING MATURATION WHICH AFFECT SEED GERMINATION

One of the most important survival mechanisms for plants growing in unpredictable conditions is the ability of one plant to produce a population of seeds with considerable variation in their germinability (heteroblasty). This will ensure that, even under optimal conditions, only part of the seed population will germinate at one time. Factors such as daylength, temperature, light quality and intensity, elevation, the mother plant's age, and seed position in the fruit, in the inflorescence, and in different inflorescences can affect heteroblasty during maturation.

Table 4
THE INFLUENCE OF THE CHANGE OF DAYLENGTH DURING THE LAST 8 DAYS OF MATURATION OF *PORTULACA OLERACEA* SEEDS, FROM 16-H DAYS TO EITHER 13- OR 8-H DAYS ON GERMINATION AT 40°C IN THE DARK WITH 5 MIN OF WHITE LIGHT ONCE IN 24 H[a]

Photoperiodic treatment of the mother plants (h) (last 8 days)	Germination percentages (%) for different daylengths			
	24 h	96 h	120 h	312 h
16	0	0	8.7 ± 4.3	88.0 ± 2.4
13	0.7 ± 0.8[b]	7.3 ± 2.9	28.0 ± 2.4	90.0 ± 1.4
8	1.3 ± 1.6	27.3 ± 4.3	48.0 ± 5.1	94.0 ± 1.4

Note: The first illumination was given 1½ h after wetting (four batches of 50 seeds each per treatment).

[a] Adapted from Gutterman.[3,4]
[b] Standard error.

A. Influences of Daylength

Portulaca oleracea seeds which matured under SD germinate faster and to a higher percentage than seeds that matured under LD (Table 4).[3,4] The critical period is the last 8 days of maturation on the mother plant. This means that even on the same plant, and even along the same branch, seeds will be produced that differ in their germinability according to the date (daylength) of the last stage of seed maturation.

Nearly the same phenomenon was found in *Polypogon monspeliensis*. Plants were grown in photoperiods of 9.0, 11.0, 12.0, 13.5, 15.0, and 18.0 h of light per day under outdoor and greenhouse conditions until the seeds matured. The plants produced seeds that germinated faster and to a higher level as the daylength increased — a facultative daylength effect on germination. The longer the daylength, the faster the germination and the higher the germination percentage. The seeds collected from plants grown under outdoor conditions germinate better than those given comparable greenhouse treatment (Table 5).[17] The same facultative effect was also found in the desert annual plant *Carrichtera annua* (Cruciferae).[17,33,34]

In contrast, both SD and LD effects were found in *Trigonella arabica*.[35] Under SD (8 h), the seeds produced by this plant are either brown or green. Under LD (20 h), seeds are yellow or yellow with green spots. Above a certain daylength (more than a specific number of hours of light per day), seeds that mature are typical of LD-treated seeds. Within 3 weeks of imbibition, 4% of yellow seeds swell and 30% of the yellow-with-green-spots seeds swell. Below a certain daylength, the seed population contains green or brown seeds of which 92 to 98% swell within 3 weeks of imbibition, respectively. All these changes in the germinability are influenced by the degree of development of the seed coat, and, thus, the color of the seed could be an indicator of the degree of impermeability of the seed coat to water. The daylength has no effect on the embryo because scarification of newly harvested seeds from all four groups will cause 100% swelling within 24 h.[35] Nearly the same phenomena were found in *Ononis sicula*, a desert annual plant of the Papilionaceae.[34]

It is interesting to note that this daylength effect does not act directly upon the seed or on the pod of *T. arabica*. The critical time period for the photoperiodic effect on germinability is the last week of seed maturation on the mother plant. Seeds matured under SD during the last 8 days of maturation, in covered as well as uncovered full-grown green pods, have the same 100% germinability. While under LD, the germinability in both cases is 10 to 30%.[35] The same phenomenon was also found in the desert annual plant *C. annua*.[16,17,33]

Table 5
THE INFLUENCE OF DIFFERENT DAYLENGTHS APPLIED FROM DECEMBER 20—24, 1979, TO *POLYPOGON MONSPELIENSIS* PLANTS DURING GROWTH AND SEED MATURATION IN THE GREENHOUSE AND OUTDOORS ON SEED GERMINABILITY

Daylength during growth of mother plant and seed maturation (h)	Germination of seeds from greenhouse plants (%)		Germination of seeds from outdoor plants (%)	
	3 days	7 days	3 days	7 days
9.0	0.0	0.0	1.5	10.0
11.0	0.0	0.0	9.5	17.0
12.0	17.5	19.0	64.0	86.5
13.5	38.0	44.5	98.5	98.5
15.0	60.5	66.0	97.0	98.5
18.0	90.5	91.0	98.0	99.0
Control natural daylength	93.0	96.0	91.5	93.0

Note: Seeds were harvested in March 1980 and germinated (4 × 50 seeds) at 25°C in continuous light in October 1980.

From Gutterman, Y., *The Physiology and Biochemistry of Seed Development, Dormancy, and Germination*, Khan, A. A., Ed., Elsevier, Amsterdam, 1982, 67. With permission.

Daylength effects on seed germination were also found in cultivated plants. The covered and uncovered fruit of tomato (*Lycopersicum esculentum*) plants grown under SD (6 h) resulted in seeds with 34 to 50% germination. Those grown under LD (20 h) resulted in seeds with 8 to 11% germination.[36] A photoperiodic effect on germination was shown in harvested fruits of cucumber (*Cucumis sativus*) that were treated for 5 short days (8 h) and another group received 5 LD (20 h). This is discussed further in Section B on hormones.

Another plant in which daylength influences seed coat development and germinability is *Chenopodium album.*[37] When the plants develop under continuous SD conditions (SSS), the harvested seeds have a seed coat thickness of 13.8 μm and a 100-seed weight of 118 mg. However, plants that develop in continuous long days (LLL) produce seeds that weigh only 66.5 mg per 100 seeds, with a seed-coat thickness of 36.8 μm. Seeds that mature in continuous long-day conditions have 24% germination in light and 3% in darkness. Seeds collected from plants grown under SSS have 100% germination in both light and dark conditions.

Plants in full flower that are transferred from long days to short days (LLS) produce seeds with an average seed-coat thickness of 29.0 μm, a 100-seed weight of 75.6 mg, and a germination percentage of 78% in light and 15% in the dark. Plants that are transferred from LD to SD in the flower bud stage for 1 month and are returned to LD in full flower (LSL), or those that remain in SD from the flower bud stage until full maturation (LSS), develop seeds with seed-coat thicknesses of 24.4 and 15.0 μm, 100-seed weights of 97 and 112.3 mg, and germination percentages of 85 and 98% in the light and 72 and 89% in the dark, respectively. Seeds that developed under continuous SD are the heaviest, have the highest germination percentages under both light and dark conditions, and have the thinnest seed coat. In contrast, seeds developed under LD have the lowest germination percentage, the lowest weight, and the greatest seed coat thickness. The photoperiodic effect on the seed germination of *C. album* was previously investigated by Wentland.[38] *C. amaranti-color* was studied by Lona,[39] and *Chenopodium* spp. by Cumming.[40]

Nearly the same results that were found by Karssen[37] in *C. album* were found by Jacques[41,42] in *C. polyspermum*, i.e., the longer the daylength during maturation in a range from 10 to 24 h daylight per day, the poorer the germination and the lower the seed weight. In another experiment on *C. polyspermum*, seeds matured in days with a 6-h photoperiod had lower germination (56%) than those produced in 9 h days (84%).[43] However, the poorest germination appeared again in seeds from 16-h days (18%), as was found by Jacques.[41,42]

Amaranthus retroflexus, a facultative SD plant for flowering, was found by Kigel et al.[44,45] to be affected by the parental photoperiodic conditions. Seeds matured under SD (8 h) germinated better than seeds matured under LD (16 h).

The germinability of seeds is also affected by 1 to 3 inductive short days before the plant flowers. In the desert annual plant *Aegilops kotschyi* the basal and upper caryopses (grains) matured under SD were found to germinate better than the basal and upper caryopses matured under LD, the daylength effect being exerted after anthesis.[46]

In contrast to what was mentioned about *Portulaca oleracea, Trigonella arabica, Carrichtera annua,* and *Ononis sicula,* transferring *Chenopodium album* from SD to LD at full flowering (LSL) did not result in production of seeds typical of those from a plant grown under continuous LD (LLL). It seems that in this case, the critical period starts much earlier than the last week of seed maturation. Similarly, *Amaranthus retroflexus* plants, induced to flower in long days by 2 to 3 short days, produced seeds that were different from those grown in continuous SD or continuous LD. In this case, the germinability is affected even before flowering, or after anthesis, as in *Aegilops kotschyi.*

B. Daylengths and Hormonal Applications

Results from hormonal applications to flower buds of *Lactuca scariola* and additional evidence about the maternal and environmental factors (daylength, etc.) affecting seed size and germination suggest that plant performance is mediated by the level of growth-regulating substances within the plant. Furthermore, the behavior of seeds and even the time of the flowering of the next generation can be modified by the parental environment via the accumulation of different levels of certain growth factors in the mature seeds (Table 6).[30]

In the cultivated plant *Cucumis sativus* different daylengths influence seed germinability of harvested ripe fruit after 5 days of storage under SD (8 h) or LD (20 h).[36] This daylength effect exists even after 270 days of seed storage. High germination (84%) resulted from SD treatments and only 12.5% germination from LD treatments. If ethephon was injected into the fruit under those photoperiodic treatments, the opposite results were obtained: low germination (2%) of seeds from SD and high germination (79%) from LD. An increase in a specific growth substance in the parent plant, or a change in the position of a certain seed (position effect), can result in changes in the relative accumulations of different growth substances in the seed, thus, affecting heteroblasty[47] (see Section IV.H).

C. Temperature Effects

A temperature effect was found in *Chenopodium album* plants grown in photoperiods of continuous LD (LLL: 18 h) or transferred to SD (8 h) after flower bud formation (LSS). Seeds matured on plants grown under temperatures of 22/12°C germinated better in either light or dark in comparison to seeds from plants grown under continuous temperatures of 22°C (Table 7).[37]

A temperature effect on seed germination and seed weight was found in *Amaranthus retroflexus*. Seeds matured on plants under continuous LD (16 h) or transferred from LD to

Tables 6

INFLUENCES OF HORMONES AND DAYLENGTH DURING FLOWERING ON *LACTUCA SCARIOLA* PLANTS (A), ON THE GERMINABILITY (B), AND ON THE TIME OF FLOWER BUD DEVELOPMENT OF THE PROGENY (C)

Table 6A: *Effects of weekly applications of growth substances* at stated times after the beginning of the differential treatments on internodes and leaf length. The chemicals were introduced by cotton threads into the main stem below the main inflorescence (ABA, GA_3) or applied as a soil drench (CCC). Means of 24 plants.

Table 6B: *Germination* (%) at 26°C either in incandescent light or in the dark of seeds matured for 16—21 days. Tests begun 14 days after the last harvest of either period. Means of 4 replicates of 50 seeds (± SE). After 20 h from the beginning of imbibition hardly any more seeds germinated either in the dark or in incandescent light. Upon transfer to red light, however, germination rose to nearly 100%, irrespective of the growth substance or daylength treatment.

Table 6C: *Flower bud development* of seedlings from seeds of similar weight harvested 81—125 days after the beginning of the differential treatments of their parent plants. 45—52 seedlings from each treatment. Values denote % of seedlings showing flower buds. Unreplicated experiment. (Adapted from Gutterman, Y., Thomas, T. H., and Heydecker, W., *Physiol. Plant.*, 34, 30, 1975.)

Table 6A
TREATED PLANTS (21 DAYS)

Treatment			Maximal internode length on main stem	Maximal leaf length above hormone supply and below inflorescence
Chemical	mg/ℓ	Daylength (h)	(mm)	(mm)
ABA	5	16	45	60
ABA	20	16	55	150
GA_3	10	16	75	55
GA_3	100	16	70	65
CCC	5000	16	11	75
Control		16	30	70
Control		16→8	30	80
Control		8	—[a]	150

Table 6B
SEED COLLECTION 16—21 DAYS FROM THE BEGINNING OF THE DIFFERENTIAL TREATMENTS

Germination (%)		
Incandescent light		Dark
12 h	20 h	20 h[b]
35 ± 5	72 ± 10	73 ± 7
52 ± 4	74 ± 2	80 ± 6
55 ± 13	80 ± 4	67 ± 9
8 ± 2	24 ± 6	39 ± 9
23 ± 3	54 ± 6	55 ± 9
46 ± 2	74 ± 0	64 ± 2
60[c]	68[c]	68[c]

Table 6C
FLOWER BUD DEVELOPMENT

Days after seedling emergence			
53	68	79	86
0	6.9	29.4	50.0
0	4.8	7.7	25.6
0	3.3	4.0	12.0
0	2.4	14.3	25.7
6.3	34.4	42.3	57.7
0	8.1	12.1	45.4
5.1	14.4	28.1	56.2

[a] No elongated internodes and no inflorescence after one year's growth.
[b] Insufficient seeds for 12 h dark.
[c] 25 seeds.

Table 7
THE EFFECT ON SEED GERMINABILITY OF *CHENOPODIUM ALBUM* OF A DAILY TEMPERATURE SHIFT (22/12°C) OR A CONTINUOUS TEMPERATURE (22°C) DURING THE GROWTH OF THE MOTHER PLANT IN CONTINUOUS LONG DAYS (LLL: 18 H) OR TRANSFERRED FROM LONG DAYS TO SHORT DAYS (8 H) AT FLOWER BUD FORMATION (LSS)

Mother plant treatment		Germination percentage (%)	
Temperature (°C)	Photoperiodic conditions	Light	Dark
22/12	LSS	100	96
	LLL	90	42
22/22	LSS	85	61
	LLL	71	22

Note: Germination was tested in continuous light or continuous darkness at 23°C.

Adapted from Karssen, C. M., *Acta Bot. Neerl.*, 19, 81, 1970. With permission.

SD (8 h) germinated to higher percentages if they were matured at temperatures of 27/22°C as compared to those matured at 22/17°C. Seeds matured at 22/17°C were found to be heavier (Table 8).[45]

A temperature effect during maturation was found also in *Aegilops ovata* (Table 9, Figure 6).[48,49] The grains were heavier when matured under 15/10°C in comparison to temperatures of 28/22°C, but the germination was much higher in the grains matured under 28/22°C (see details in Section IV.H and also Wurzberger and Koller[46]).

In addition to the daylength effect, a temperature effect on seed germinability was found during seed maturation in *P. oleracea* plants (Table 10).[3] In all cases, the percentage of seeds germinating in the dark was found to be lower in seeds matured under LD (16 h; 27/22°C) in comparison to the long (15 h) natural daylength and temperatures ranging from a maximum of 32°C to a minimum of approximately 11°C. In both seed populations there was an increase in the germination percentage with an increase in time of storage.

A difference in the speed of germination was also found.[3] The seeds mentioned above that matured under natural conditions started to germinate earlier, after 3 h of imbibition at 40°C and continuous light, whereas seeds matured under continuous LD started to germinate after 5 h of imbibition. Within 24 h the seeds from both populations reached more than 90% germination.

D. Red and Far-Red Light

In *P. oleracea* certain low-intensity red and far-red light treatments during plant development and seed maturation have had an influence on seed germinability for years. The most pronounced effect on germination in either continuous white light or continuous red light was found in seeds which were collected from plants that received 2 h of red or far-

Table 8
EFFECT OF TEMPERATURE AND PHOTOPERIODIC TREATMENTS ON SEED GERMINABILITY AND SEED WEIGHT OF *AMARANTHUS RETROFLEXUS* PLANTS DURING DEVELOPMENT AND SEED MATURATION

Mother plant treatment			
Temperature (°C)	**Photoperiodic conditions**	**Germination percentage (%)**	**Seed weight**
27/22	LD→SD	47	41.1 ± 1.0[a]
	LD	21	38.3 ± 1.0
22/17	LD→SD	18	46.6 ± 0.7
	LD	11	44.7 ± 1.0

Note: LD = continuous long days (16 h), LD → SD = plants transferred from long days (16 h) to short days (8 h) when 50% of the plants in the population were macroscopically flowering. Seed germinability tests were conducted at 30°C in the dark with pretreatments of dark incubation at 5°C for 3 days.

[a] Standard error.

Adapted from Kigel, J., Gibly, A., and Negbi, M., *J. Exp. Bot.*, 30, 997, 1979. With permission.

red light after 8 h of daylight followed by 14 h of darkness (Figure 1). When 8 h of daylight are followed by 8 h of red or far-red light during maturation, there is no difference in the seed germination under continuous white light. However, under red light conditions a difference in germination appears which may have been induced during seed maturation.

In *P. oleracea* plants, none of the photoperiodic treatments or treatments of red and far-red light during maturation of the seeds on the mother plants, including the treatment of continuous red light during the last 8 days of seed maturation, brought about germination of the seeds in continuous darkness.[3,4]

This is not the case with *Arabidopsis thaliana.* Continuous red light applied during seed maturation causes germination of the seeds in the dark.[50] Post-harvest red and far-red light treatments on *Cucumis sativus* fruit influence the germinability of the seeds when germinated immediately before dehydration (about 45% seed water content). No effect was evident after dehydration and rehydration.[36,51] In *Chenopodium album,* the red light effect disappears 4 months after seed maturation.[37]

At this point, the general question raised is — What is the nature of the phytochrome in the embryo before dehydration, in fixation during dehydration, and during rehydration in different species after different times of storage?

The influences of the red/far-red light may have some ecological implications on the germination of light-sensitive seeds under a leaf canopy. Leaves act as a far-red filter and can be used to demonstrate that far-red light inhibits seed germination (see Chapter 2 for further discussion).[52,53] Not only the leaf canopy, but also the investing structures could influence seed germination. As was found in *C. album* and other arable weeds, the maternal tissues that surround the developing seeds retain high concentrations of chlorophyll, and during seed maturation and desiccation act as a far-red filter, prevent the germination of seed in the dark.[54] It is also possible that in *P. oleracea* the chlorophyll remaining in the

Table 9
POSITION EFFECT ON AVERAGE WEIGHT (MG) AND GERMINATION (%) AFTER 24 H OF CONTINUOUS LIGHT AT 15°C ON *AEGILOPS OVATA* CARYOPSES, GROWN AT TEMPERATURES OF 15°C DURING THE DAY AND 10°C DURING THE NIGHT, OR AT TEMPERATURES OF 28°C DURING THE DAY AND 22°C DURING THE NIGHT, BOTH GROUPS GROWN UNDER LONG-DAY CONDITIONS (18 H)

Order of caryopses from which mother plant developed[a]	Order of caryopses collected from mother plants	Average weight of caryopses (mg)		Germination (%)	
		15/10°C	28/22°C	15/10°C	28/22°C
	a_1	20.6 ± 0.7[b]	13.9 ± 0.5	84.4	100
a_1	b_2	9.5 ± 0.8	6.7 ± 0.3	10.0	60
	$c_{(2)}$	6.1 ± 0.6	3.0 ± 0.3	8.5	63.1
	a_1	22.9 ± 0.5	12.9 ± 0.4	55.0	100
b_2	b_2	9.2 ± 0.4	7.1 ± 0.2	2.3	85.3
	$c_{(2)}$	3.8 ± 0.4	3.0 ± 0.2	0	90
	a_1	27.3 ± 1.9	14.7 ± 1.7	21.2	100
c	b_2	13.0 ± 0.3	6.9 ± 0.8	0	76
	$c_{(2)}$	4.2 ± 0.7	3.5 ± 0.4	0	87

[a] In spikes containing three spikelets, there are usually five different grains. In the lowest spikelet there are two grains: a large yellow grain (a_1) and a small brown one (a_2); in the second spikelet, there are two grains: a yellow grain smaller than a_1 (b_1) and another brown one smaller than a_2, (b_2). The terminal spikelet contains only one grain, $c_{(2)}$, which resembles a_2 and b_2. (See Figure 6.)

[b] Standard error.

Adapted from Datta, J. C., Gutterman, Y., and Evenari, M., *Planta*, 105, 155, 1972. With permission.

fruit, which keeps the fruit green almost until full maturation, has an effect on the light sensitivity of the seeds. This phenomenon has a very important ecological implication. The small seeds of *Portulaca* and *Chenopodium* are, thus, prevented from germinating at soil depths in which light penetration is not sufficient for the conversion of Pr to Pfr.

A similar effect could be caused by leaf shadow on the developing seed during maturation. This canopy effect is an additional factor that influences heteroblasty. This is very important for weed survival under unpredictable conditions. The canopy effect could also prevent germination of light-sensitive seeds under conditions of high competition.

E. Light Intensity

In *Amaranthus retroflexus*, seed germination is affected by the light intensity under which the mother plant develops and matures.[44] The germination of seeds in the dark that were matured under SD and shade conditions was higher than that of seeds matured under SD and full light intensity. Seeds collected from plants that developed under LD of full light intensity germinated better, in most cases, than seeds from plants exposed to long days but placed in the shade.

In *Datura ferox*, a summer weed, a low-light intensity as well as water stress during seed maturation increased the germination percentages and decreased the level of seed germination inhibitors.[55]

F. Altitude Effects on Germination

Seeds of *C. bonus-henricus* plants that were collected from altitudes of 600 to 2600 m

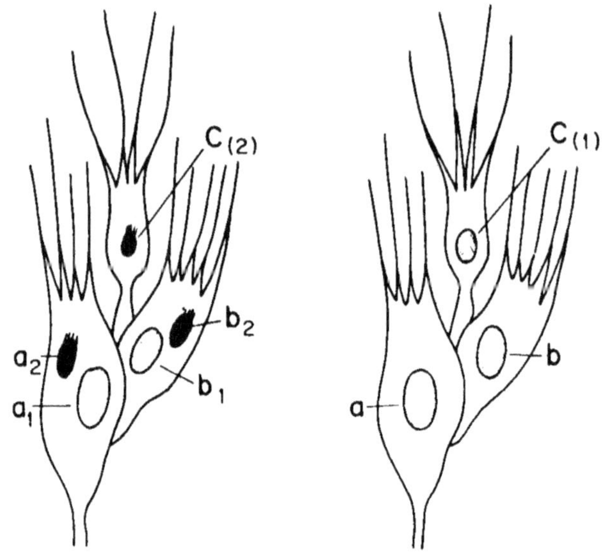

FIGURE 6. Schematic drawing of two spikes of type III of *Aegilops ovata*. On the left spike, the two lower spikelets contain two caryopses each (a_1, a_2 and b_1, b_2). The higher spikelet contains only one caryopsis ($c_{(2)}$). The a_1 and b_1 caryopses are larger, heavier, brighter, and less hairy than the a_2, b_2, and $c_{(2)}$. On the right spike, each spikelet contains only one caryopsis (a, b, and $c_{(1)}$). All of them are similar to caryopses a_1 and b_1. (From Datta, J. C., Evenari, M., and Gutterman, Y., *Isr. J. Bot.*, 19, 463, 1970. With permission.)

showed large differences in germination percentages.[56] The higher the altitude, the lower the amount of germination, the thicker the seed coat, and the greater the content of polyphenols in the seed. Progenies of plants transferred from high to low elevation and from low to high elevation showed a direct influence of the new environment on the seed coat inhibition of germination. The polyphenols accumulated in the thick seed coat when the plants were placed in a high elevation inhibit germination. The increase in polyphenol content may be due to increased visible radiation as the elevation increases.

G. Senescence Effects

Seeds of *Trigonella arabica* were found to have incomplete seeds coats, typical to maturation under SD, as well as when matured under LD and taken from plants that had started to dry out at the end of the season.[57] At this stage, the pods had reached their final size and were still green. This same phenomenon was also found in *Ononis sicula*.[57]

H. Position Effects

Seeds that mature in different inflorescences, in different locations in the same inflorescence, or even in different parts of the same fruit may not have the same germinability. Position effects on germination ensure heteroblasty and are important for the survival of weeds.

An example of differences in the germination between inflorescences can be seen in *Emex spinosa*, an amphicarpic annual plant that produces aerial, telechoric, and subterranean atelechoric propagules. The aerial propagules are spiney, small, and light. The subterranean propagules are smooth, large, and heavy. They differ in their germinability. The aerials contain germination inhibitors and, therefore, leaching improves their germination. The subterranean propagules have no germination inhibitors and leaching only minimally im-

Table 10
PORTULACA OLERACEA SEED GERMINATION AS A FUNCTION OF TEMPERATURE DURING MATURATION

Date of germination test	Maturation conditions (°C)	Germination percentages (%) at different temperatures (°C)						
		10	20	25	30	35	40	45
August 16, 1967	27/22	0	0	0	7.5 ± 0.8[a]	0	6.0 ± 1.4	1.5 ± 0.8
	32—11	0	0	0	13.5 ± 0.8	2.5 ± 1.6	8.0 ± 1.4	8.0 ± 2.8
January 21, 1968	27/22	0	0	0	0	2.0 ± 0.9	14.0 ± 2.1	31.5 ± 4.9
	32—11	0	7.5 ± 2.0	7.5 ± 0.6	20.0 ± 3.2	31.0 ± 3.9	41.5 ± 1.1	53.5 ± 6.6
November 5, 1968	27/22	0	0	1.0 ± 1.1	6.5 ± 2.2	13.5 ± 3.9	44.0 ± 0.9	41.0 ± 4.8
	32—11	0	4.0 ± 0.9	6.5 ± 2.0	27.5 ± 2.0	40.5 ± 4.7	60.0 ± 4.9	55.0 ± 2.0

Note: Seeds matured under long days at 27/22°C and at natural temperatures (max approx. 32°C, min approx. 11°C) were collected in June 1967. Results are taken from three out of a set of eight germination tests carried out between August 16, 1967 to November 5, 1968. Seeds were germinated for 24 h in continuous dark at constant temperatures from 10—45°C.

[a] Standard error.

Adapted from Gutterman, Y., The Photoperiodic Response of Some Plants and the Effect of the Environment of the Mother Plants on the Germination of the Seeds, Ph.D. thesis, Hebrew University of Jerusalem, 1969. With permission.

proves their germination. Cutting open the micropylar tips in the aerial propagules causes approximately 80% germination in a temperature range of 10 to 30°C in the light or in the dark. In subterranean seeds, light inhibits germination when temperatures range from 5 to 15°C, but not in temperatures from 20 to 25°C (Table 11).[58] Germination differences between achenes matured on aerial capitula (inflorescence) and those on subterranean ones were also found in the desert annual amphicarpic plant *Gymnarrhena micrantha*.[59]

In the cultivated plant *Apium graveolens* (celery), different inflorescences (primary, secondary, tertiary, and quarternary umbels) produce seeds of different weights and different germinability.[60] A comparison between the germination of seeds of the same size from different umbel orders in *Daucus carota* (carrot) showed that seeds from primary umbels had a higher germination percentage than those of the same size from secondary and tertiary orders. In this case, a clear position effect was shown.[61]

An example of a position effect on germination in a spike (inflorescence) dispersal unit was found in *Aegilops ovata*. The spike is composed of two to four spikelets (a, b, c, d).[48,49] The lower ones contain two caryopses with grains a_1, a_2, and b_1, b_2 (Figure 6). The upper ones contain a single caryopsis each, c and d. The lower grains of each spikelet (a_1 and b_1) are larger than a_2 and b_2, whereas the smallest caryopses are in spikelets c and d. Under natural conditions only the a_1 grain usually germinates from the whole unit during the first year. The heteromorphous grain population of one dispersal unit is also heteroblastic.[48]

Plants originating from caryopses of the diverse orders (a_1, b_2, c) from spikes containing three spikelets were grown under thermoperiods of 28/22°C and 15/10°C in LD of 18 h.[48,49] When mother plants arose from various orders of caryopses and were grown under LD and temperatures of 15/10°C, they produced relatively heavier caryopses of different germinability and slower germination than did corresponding plants grown at 28/22°C (Table 9). Mother plants grown from grains a_1, b_2, and c under temperatures of 15/10°C developed, in turn, grains with different germinability. The a_1 grains that developed on mother plants grown from a_1 grains had the highest germination percentage. Grains with the worst germination were from mother plants developed from c grains. Under temperatures of 28/22°C there is no such effect on germination. There is a pronounced position effect on germination from each harvested spikelet grown under 15/10°C, but this effect is relatively small in spikes matured in temperatures of 28/22°C (Table 9).[48,49]

These results show that the origin of the mother plant may affect the germination quality of its offspring. In this case, certain characteristics that are, to some degree, dependent on the position and germination behavior of the seeds of the first generation are transmitted through the second generation to the following generation or generations. Wurzberger and Koller[46] found a similar phenomenon in *Aeglops kotschyi*.

Another example of an inflorescence as a dispersal unit having a position effect on germination was found in the desert annual plant *Pteranthus dichotomus*. In this plant the terminal third-order seeds germinate better than the second order. The first-order seeds have the lowest percentage of germination.[62]

As was summarized in the detailed paper of Grey and Thomas,[47] a position effect on germination and/or on subsequent growth of the seeds is reported for several cultivated plants and weeds such as rice, maize, wheat, oats, cotton, cabbage, onion, cacao, eggplant, lettuce, *Rumex crispus*, carrots, and *Atriplex hortensis*. A position effect on the germination of seeds from one fruit (capsule) was found in the desert ombrohydrochoric (dispersed by rain) annual plant *Mesembryanthemum nodiflorum*.[63] This fruit contained about 30 seeds. The terminal third group (approximately one third of the seeds) in the capsule dispersed after 15 min of wetting and had the highest germination, about 60%. The middle group of seeds were dispersed by rain after 200 min of wetting and germinated about 5 to 10%. The bottom third were dispersed after 320 min of wetting and had the poorest germination. There is no difference in the size and weight of the seeds of this plant.[63]

Table 11
GERMINATION PERCENTAGES OF LEACHED AND UNLEACHED *EMEX SPINOSA* AERIAL AND SUBTERRANEAN SEEDS AT VARIOUS TEMPERATURES IN LIGHT (L) AND DARK (D) CONDITIONS

Temperature (°C)	Percentage germination (%)							
	Aerial seeds						Subterranean seeds	
	Unleached		Leached		Unleached with micropylar tip cut open		Unleached with micropylar tip cut open	
	L	D	L	D	L	D	L	D
5	0	0	0	0	78 ± 2	75 ± 3	18 ± 2	68 ± 1
10	0	0	5 ± 0.8	13 ± 1.1	80 ± 3	78 ± 3	24 ± 1	74 ± 3
15	2 ± 1	5 ± 1	20 ± 1.7	44 ± 1.8	84 ± 2	82 ± 3	40 ± 2	80 ± 4
20	7.5 ± 1.5	24 ± 1.5	36 ± 1.5	60 ± 2.0	85 ± 1	85 ± 1	80 ± 1	85 ± 3
25	7 ± 2.0	3 ± 1.5	36 ± 2.0	41 ± 1.2	85 ± 2	85 ± 3	80 ± 4	80 ± 1
30	0	0	30 ± 1.9	17 ± 1.9	80 ± 3	80 ± 4	70 ± 2	80 ± 3
35	0	0	10 ± 2.0	5 ± 0.7	—	—	—	—

Adapted from Evenari, M., Kadouri, A., and Gutterman, Y., *Compd. Flora*, 166, 223, 1977. With permission.

I. Conclusions

Embryonic tissues in higher annual plants undergo changes during maturation and dehydration, enabling the seeds to retain their ability to germinate and develop a new plant over relatively long time periods. The seed is the only organ in the higher annual plants that has this ability. The seed makes it possible for these plants to live in an area that is suitable only a few months of the year for their growth and development. This special ability for the seed to survive as a dormant tissue for long periods and germinate at the right time (''distribution in time'') is a very important survival advantage for weeds and also for other plants that inhabit unpredictable environments, such as deserts, with extreme hot and cold temperatures.

The seeds accumulate and disperse in the soil depths through soil cracks, by soil movement, and plowing. In these ways, they can be distributed to new habitats (''distribution in space'').

Environmental and maternal factors affecting seed maturation during plant development can influence germination by acting on the seed coat and the embryo itself. It was found that environmental factors influence seed coat thickness in the following plants: *C. amaranticolor,*[39] *C. polyspermum,*[42] and *C. album.*[37] In *C. bonus-henricus,* elevation and increased polyphenol content may inhibit germination.[56]

Other ways in which the seed coat affects germinability could be through changes in the coat's impermeability to water (see Chapter 2 for a detailed discussion). In *T. arabica*[33,35] and *O. sicula,*[34] the formation of the seed coat may be regulated by enzymes related to the synthesis of phenolic-type compounds, e.g., polyphenol oxidase, as was found in another species of the Papilionaceae, *Pisum elatius,* a wild strain of pea.[64,65] During the development of the water-impermeable seed coats in *Sida spinosa,* it seems that peroxidase is involved in the polymerization of soluble phenolics to insoluble lignin polymers.[66]

The influence of environmental factors[67,68] that may bring about changes in the embryo, endosperm, enzymes, hormones, and other materials possibly involved in the germination process were observed in the plants *Portulaca oleracea,*[4] *Polypogon monspeliensis,*[17] *Carrichtera annua,*[17,33,34] *Amaranthus retroflexus,*[44,45] and *Chenopodium album* (in addition to the seed coat effect).[37] In addition, the position of the seeds in the fruit, the inflorescence, or in different inflorescences usually affect the size and weight of the seeds as well. This was shown in *Emex spinosa,*[58] *Gymnarrhena micrantha,*[59] *Apium graveolens,*[60] *Aegilops ovata,*[48,49] *Pteranthus dichotomus,*[62] and *Mesembryanthemum nodiflorum.*[63] A direct phytochrome effect during maturation on germination was shown in *Arabidopsis thaliana,*[50] *Cucumis sativus,*[51] *Chenopodium album,*[37,54] and *Portulaca oleracea.*[4]

From the above one can see that the influences during maturation on seed germination could be of great value as a survival mechanism of annual plants, producing seeds with different germinability on the same plant. This is very important to the survival of plants such as weeds and desert plants that grow under unpredictable conditions.

V. SUMMARY

Reviewed in this chapter were the flowering mechanisms of annual weeds, seed development, germinability of the seeds, and the influence of the environment on seed maturation while the seed is still on the mother plant.

It is interesting to note that almost all of the prevalent weed species are able to flower within a short time after germination and either are not influenced at all by daylength or can flower under a wide range of daylengths. Ecotypes of one species in different geographical areas were found which flower under a variety of daylengths, specific to the season of the year which provides the optimal conditions for each type. It was found on one occasion that summer and winter cytological races alternated in the same area. This flexibility of some annual weed species enables them to spread to areas all over the world.

The process of seed maturation is influenced by the environment and conditions in the mother plant. Seeds on the same plant, on the same branch, or even within the same fruit may differ in their germinability. These differences, which are determined while the seeds are still on the mother plant, ensure a distribution over time in germination. Even under optimal conditions only a part of the seeds germinate at one time, while the rest remain in the earth's seed bank.

The light requirement for germination ensures that small seeds will germinate only when on or near the soil surface. This important survival mechanism is also determined, at least in some species, while the seeds are still on the mother plant.

These mechanisms and others that are mentioned in subsequent chapters help us to understand better the tenacious survival of weeds, despite man's efforts to destroy them.

ACKNOWLEDGMENTS

The author wishes to acknowledge Dr. C. M. Karssen, Dr. J. Kigel, Dr. S. O. Duke, Dr. G. H. Egley, Prof. A. A. Khan, and Dr. T. H. Thomas who sent him manuscripts of their final studies and publications which enabled him to provide up-to-date research data in this chapter.

REFERENCES

1. **Holm, L. G., Plucknett, D. L., Pancho, J. V., and Herberger, J. P.,** *The World's Worst Weeds: Distribution and Biology*, University Press of Hawaii, Honolulu, 1977.
2. **Kigel, J. and Rubin, B.,** *Portulaca oleracea*, in *Handbook on Flowering*, Vol. 4, Halevy, A., Ed., CRC Press, Boca Raton, Fla., 1985.
3. **Gutterman, Y.,** The Photoperiodic Response of Some Plants and the Effect of the Environment of the Mother Plants on the Germination of the Seeds, Ph.D. thesis, Hebrew University of Jerusalem, 1969.
4. **Gutterman, Y.,** Influences of the photoperiodic regime and red-far red light treatments of *Portulaca oleracea* L. plants on the germinability of their seeds, *Oecologia*, 17, 27, 1974.
5. **Kigel, J.,** *Amaranthus*, in *Handbook on Flowering*, Vol. 1, Halevy, A., Ed., CRC Press, Boca Raton, Fla., 1985.
6. **Koller, D., Kigel, K., Nir, I., and Ofir, M.,** Environmental Control of Weed Physiology, *Final Report U.S.D.A. Project A10-CR-27*, U.S. Department of Agriculture, Washington, D.C., 1975.
7. **Kigel, J.,** *Sorghum halepense* L., in *Handbook of Flowering*, Vol. 4, Halevy, A., Ed., CRC Press, Boca Raton, Fla., 1985.
8. **McMillan, C.,** Photoperiod in three *Xanthium* populations from the Tropic of Cancer in Mexico, *Science*, 165, 292, 1969.
9. **Ray, P. M. and Alexander, W. E.,** Photoperiodic adaptation to latitude in *Xanthium strumarium*, *Am. J. Bot.*, 53, 806, 1966.
10. **Cumming, B. G.,** Early-flowering plants, in *Methods in Developmental Biology*, Wilt, F. and Wessells, N., Eds., Thomas Y. Crowell, New York, 1967, 277.
11. **Ramakrishnan, P. S. and Kapoor, P.,** Photoperiodic requirements of seasonal populations of *Chenopodium album* L., *J. Ecol.*, 62, 67, 1974.
12. **Kaul, V.,** Physiological-ecology of *Xanthium strumarium* Linn. IV. Effect of climatic factors on growth and distribution, *New Phytol.*, 70, 799, 1971.
13. **Williams, R. D.,** Photoperiod effects on the reproductive biology of purple nutsedge (*Cyperus rotundus*), *Weed Sci.*, 26, 539, 1978.
14. **Evenari, M. and Gutterman, Y.,** The photoperiodic response of some desert plants, *Z. Pflanzenphysiol.*, 54, 7, 1966.
15. **Evenari, M. and Gutterman, Y.,** Desert plants, in *Handbook of Flowering*, Vol. 1, Halevy, A. H., Ed., CRC Press, Boca Raton, Fla., 1985.
16. **Gutterman, Y.,** Review: survival mechanisms of desert winter annual plants in the Negev Highlands of Israel, in *Scientific Reviews on Arid Zone Research*, Mann, H. S., Ed., Scientific Publishers, Jodhpur, India, 1982, 249.

17. **Gutterman, Y.**, Phenotypic maternal effect of photoperiod on seed germination, in *The Physiology and Biochemistry of Seed Development, Dormancy and Germination*, Khan, A. A., Ed., Elsevier, New York, 1982, 67.
18. **Bewley, J. D. and Black, M.**, *Physiology and Biochemistry of Seeds in Relation to Germination, Development, Germination and Growth*, Springer-Verlag, New York, 1978, 40.
19. **Duffus, C. M. and Cochrance, M. P.**, Carbohydrate metabolism during seed grain development, in *The Physiology and Biochemistry of Seed Development, Dormancy and Germination*, Khan, A. A., Ed., Elsevier, New York, 1982, 43.
20. **Khan, A. A.**, Gibberellins and seed development, in *The Physiology and Biochemistry of Seed Development, Dormancy, and Germination*, Khan, A. A., Ed., Elsevier, New York, 1982, 111.
21. **King, R. W.**, Abscisic acid in seed development, in *The Physiology and Biochemistry of Seed Development, Dormancy and Germination*, Khan, A. A., Ed., Elsevier, New York, 1982, 157.
22. **van Staden, J., Davey, J. E., and Brown, A. C.**, Cytokinins in seed development and germination, in *The Physiology and Biochemistry of Seed Development, Dormancy and Germination*, Khan, A. A., Ed., Elsevier, New York, 1982, 137.
23. **Weidner, S. and Zalewski, K.**, Changes in ribosomal proteins in wheat embryos in course of grain development and maturation, *Acta Soc. Bot. Pol.*, 51, 283, 1982.
24. **Brown, J. W. S., Erstand, D. R., and Hall, T. C.**, Molecular aspect of storage protein synthesis during seed development, in *The Physiology and Biochemistry of Seed Development, Dormancy and Germination*, Khan, A. A., Ed., Elsevier, New York, 1982, 3.
25. **Pate, J. S.**, Pea, in *Crop Physiology*, Evans, L. T., Ed., Cambridge University Press, New York, 1975, 191.
26. **Eeuwens, C. G. and Schwabe, W. W.**, Seed and pod wall development in *Pisum sativum* L. in relation to extracted and applied hormones, *J. Exp. Bot.*, 26, 1, 1975.
27. **Bhatnagar, S. P. and Johri, B. M.**, Development of angiosperm seeds, in *Seed Biology Vol. 1*, Kozlowski, T. T., Ed., Academic Press, New York, 1972, 77.
28. **Durley, R. C., MacMillan, J., and Pryce, R. J.**, *Phytochemistry*, 10, 1891, 1971.
29. **Barendse, G. W. M. and Kende, A.**, Fate of radioactive gibberellin A_1 in maturing and germinating seeds of peas and Japanese Morning Glory, *Plant Physiol.*, 43, 815, 1968.
30. **Gutterman, Y., Thomas, T. H., and Heydecker, W.**, Effect on the progeny of applying different day length and hormone treatments to parent plants of *Lactuca scariola*, *Physiol. Plant.*, 34, 30, 1975.
31. **Zeevaart, J. A. D.**, Reduction of the gibberellin content of pharbitis seeds by CCC and after-effects in the progeny, *Plant Physiol.*, 41, 856, 1966.
32. **Felippe, G. M. and Dale, J. E.**, Effects of CCC and gibberellic acid on the progeny of treated plants, *Planta*, 80, 344, 1968.
33. **Gutterman, Y.**, Influences of environmental conditions and hormonal treatment of the mother plants during seed maturation on the germination of their seeds, in *Advances in Plant Reproductive Physiology*, Malik, C. P., Ed., Kalyani Publishers, New Delhi, 1977, 288.
34. **Gutterman, Y.**, Differences in the progeny due to daylength and hormonal treatment of the mother plants, in *Seed Ecology*, Heydecker, W., Ed., Pennsylvania State University Press, University Park, 1973, 59.
35. **Gutterman, Y.**, Seed coat permeability as a function of photoperiodical treatment of the mother plants during seed maturation in the desert annual plant *Trigonella arabica* Del., *J. Arid Environ.*, 1, 141, 1978.
36. **Gutterman, Y.**, Germinability of seeds as a function of the maternal environment, *Acta Hortic.*, 83, 49, 1978.
37. **Karssen, C. M.**, The light promoted germination of the seeds of *Chenopodium album* L., *Acta Bot. Neerl.*, 19, 81, 1970.
38. **Wentland, M. J.**, The Effect of Photoperiod on the Seed Dormancy of *Chenopodium album*, thesis, University of Wisconsin, Madison, 1965.
39. **Lona, F.**, L'Influenza della Condizioni Ambientali, Durante l'Embriogenesi, sulla Caratteristiche della Pianta che ne Deriva, Publ. Ist. Bot., University of Milano, Italia, 1947, 1.
40. **Cumming, B. G.**, The dependence of germination on photoperiod, light quality and temperature in *Chenopodium* ssp., *Can. J. Bot.*, 41, 1211, 1963.
41. **Jacques, R.**, Quelques Donnees sur le Photoperiodisme de *Chenopodium polyspermum* L.; Influence sur la Germination des Graines, Colloqua International sur le Photo-thermoperiodisma, Publ. 34, Series B, International Union of Biological Science, Utrecht, The Netherlands, 1957, 125.
42. **Jacques, R.**, Action de la lumiere par l'intermediaire du phytochrome sur la germination la croissance et le developpement de *Chenopodium polyspermum* L., *Physiol. Veg.*, 6, 137, 1968.
43. **Gutterman, Y.**, The influences of daylength during maturation on germination of *Chenopodium polyspermum* seeds, unpublished results.
44. **Kigel, J., Ofir, M., and Koller, D.**, Control of the germination responses of *Amaranthus retroflexus* L. seeds by their parental photothermal environment, *J. Exp. Bot.*, 28, 1125, 1977.

45. **Kigel, J., Gibly, A., and Negbi, M.,** Seed germination in *Amaranthus retroflexus* L. as affected by the photoperiod and age during flower induction of the parent plants, *J. Exp. Bot.*, 30, 997, 1979.
46. **Wurzberger, J. and Koller, D.,** Differential effects of the parental photothermal environment on development of dormancy in caryopses of *Aegilops kotschyi*, *J. Exp. Bot.*, 27, 43, 1976.
47. **Grey, D. and Thomas, T. H.,** Seed germination and seedling emergence as influenced by the position of development of the seed on, and chemical applications to, the parent plant, in *The Physiology and Biochemistry of Seed Development, Dormancy and Germination*, Khan, A. A., Ed., Elsevier, New York, 1982, 137.
48. **Datta, J. C., Evenari, M., and Gutterman, Y.,** The heteroblasty of *Aegilops ovata* L., *Isr. J. Bot.*, 19, 463, 1970.
49. **Datta, J. C., Gutterman, Y., and Evenari, M.,** The influence of the origin of the mother plant on yield and germination of their caryopses in *Aegilops ovata*, *Planta*, 105, 155, 1972.
50. **McCullogh, J. M. and Shropshire, W., Jr.,** Physiological predetermination of germination responses in *Arabidopsis thaliana* L., Heynh., *Plant Cell Physiol.*, 11, 139, 1970.
51. **Gutterman, Y. and Porath, D.,** Influences of photoperiodism and light treatments during fruit storage on the phytochrome and on the germination of *Cucumis prophetarum* L. and *Cucumis sativus* L. seeds, *Oecologia*, 18, 37, 1975.
52. **Black, M.,** Light-controlled germination of seeds, *Soc. Exp. Biol. Symp.*, 23, 193, 1969.
53. **VanderVeen, R.,** The importance of the red-far red antagonism in photoblastic seeds, *Acta Bot. Neerl.*, 19, 809, 1970.
54. **Cresswell, E. G. and Grime, J. P.,** Induction of a light requirement during seed development and its ecological consequences, *Nature (London)*, 291, 583, 1981.
55. **Sanchez, R. A., Eyherabide, G., and de Migual, L.,** The influence of irradiance and water deficit during fruit development on seed dormancy in *Datura ferox* L., *Weed Res.*, 21, 127, 1981.
56. **Dorne, G. J.,** Variation in seed germination inhibition of *Chenopodium bonus-henricus* in relation to altitude of plant growth, *Can. J. Bot.*, 59, 1893, 1981.
57. **Gutterman, Y.,** Review: influences on seed germination: phenotypic maternal effects during seed maturation, *Isr. J. Bot.*, 29, 1980/81.
58. **Evenari, M., Kadouri, A., and Gutterman, Y.,** Eco-physiological investigations on the amphicarpy of *Emex spinosa* L., *Compd. Flora*, 166, 223, 1977.
59. **Koller, D. and Roth, N.,** Studies on the ecological and physiological significance of amphicarpy in *Gymnarrhena micrantha* (Compositae), *Am. J. Bot.*, 51, 26, 1964.
60. **Thomas, T. H., Biddington, N. L., and O'Toole, F.,** Relationship between position on the parent plant and dormancy characteristics of seeds of three cultivars of celery (*Apium graveolens*), *Physiol. Plant*, 45, 492, 1979.
61. **Jacobsohn, R. and Globerson, D.,** *Daucus carota* (carrot) seed quality. I. Effects of seed size on germination, emergence and plant growth under subtropical conditions. II. The importance of the primary umbel in carrot seed production, in *Seed Production*, Hebblethwaite, P. D., Ed., Butterworth's, London, 1980, 637.
62. **Evenari, M.,** Zur Keimungsaekalogie zweier wuestenpflanzen. Mitt, *Florist.-Soziol. Arbeitsgem*, 10, 70, 1963.
63. **Gutterman, Y.,** Annual rhythm and position effect in the germinability of *Mesembryanthemum nodiflorum*, *Isr. J. Bot.*, 29, 93, 1980/81
64. **Marbach, I. and Mayer, A. M.,** Permeability of seed coats to water as related to drying conditions and metabolism of phenolics, *Plant Physiol.*, 54, 817, 1974.
65. **Marbach, I. and Mayer, A. M.,** Changes in catechol oxidase and permeability to water in seed coats of *Pisum elatius* during seed development and maturation, *Plant Physiol.*, 56, 93, 1975.
66. **Egley, G. H., Paul, R. N., Jr., Vaughn, K. C., and Duke, S. O.,** Role of peroxidase in the development of water-impermeable seed coats in *Sida spinosa* L., *Planta*, 157, 224, 1983.
67. **Zimmerman, C. A.,** Growth characteristics and weediness in *Portulaca oleracea* L., *Ecology*, 57, 964, 1976.
68. **Singh, K. P.,** Effect of different photoperiods on growth and flowering in *Portulaca oleraceae* L., *Curr. Sci.*, 41, 573, 1972.

Chapter 2

PHYSIOLOGY OF WEED SEED DORMANCY AND GERMINATION

Grant H. Egley and Stephen O. Duke

TABLE OF CONTENTS

I. SIGNIFICANCE OF WEED SEED DORMANCY

Dormancy is a property of many weed seeds that enables them to survive conditions hazardous to plant growth and to germinate at some later time or in some other place.[1] Because of dormancy, seeds may persist in the soil for many years and germinate when conditions are favorable for seedling survival through to maturity.[2-5] The vast numbers of weed seeds that may occur in agricultural soils provide a source of weed seedlings that will create weed problems during the period of weed seed persistence in the soil.

Because weed seeds in the soil may be in different dormancy states, germination within a seed population may be nonsynchronous, thus, spreading seedling emergence over a period of time and causing the farmer to repeatedly face similar weed problems. Harper[6] stressed the importance of dormancy to weed seed survival and he recognized the problem that dormancy poses to advances in weed control. Dormant weed seeds are difficult to kill and, with the exception of soil sterilization methods, current weed control techniques are not effective upon dormant weed seeds.[7] Improved knowledge of weed seed dormancy may lead to new and improved methods of weed control. Chancellor[8] stated that the next advance in weed control will be the reduction of dormant weed seed populations with chemical germination stimulants.

The physiology of weed seed dormancy is poorly understood. Dormancy is usually considered as the failure of a seed to germinate under conditions normally favorable for growth of the seedling.[9] Dormancy has a genetic basis but the gene expression may be influenced by environmental factors, some of which may occur while the seeds are developing on the parent plant (see Chapter 1). Seeds that are dormant immediately after maturation are said to be in primary (or innate) dormancy.[10] Seeds that become dormant at some time after maturity and seed dispersal are considered to be in secondary dormancy. Originally, non-dormant seeds and seeds that have had their primary dormancy broken may pass into secondary dormancy if their germination is inhibited under certain conditions (Figure 1). In the field, weed seeds may alternately lose and acquire dormancy and display seasonal rhythmic germinability during their period of persistence in the soil.[11] For example, seeds of a summer annual that have primary dormancy when buried in the autumn may have primary dormancy broken during the winter, and then pass into secondary dormancy during the following spring and summer months. Secondary dormancy can be broken during the next winter (Figure 2). Presumably, similar cycles occur during succeeding years.

What are the biological bases for seed dormancy? What mechanisms enable a seed to break dormancy and germinate when conditions will be favorable for germination and also for seedling growth after germination? The dormant seed must be responding to stimuli that indicate the likelihood of future conditions, otherwise seed dormancy would not necessarily provide survival benefits for the species.[12]

Dormancy is not unique to weed seeds but occurs in propagules of many wild species and some domesticated species as well. Most domesticated species have had their undesirable seed dormancies removed through selection and breeding so that predictable, synchronous germination can occur. Our discussions will emphasize weed seeds but will include information on dormancy and germination of other seeds for illustration of important points.

We consider dormancy as the failure of a viable seed to germinate under conditions favorable for growth of the seedling. We describe germination as the resumption of embryo growth sufficient to result in protrusion of a portion of the embryo through all coverings over the embryo. Post-germination events such as seedling growth after embryo protrusion and conversion of storage foods to sustain seedling growth will not be included in the discussions.

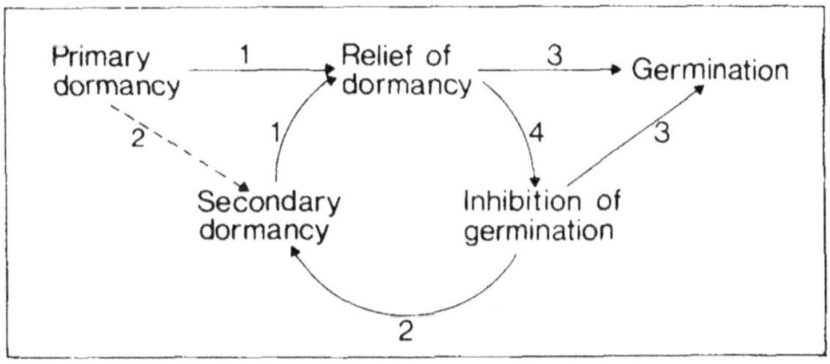

FIGURE 1. Dormancy cycle in seeds. Dormancy changes caused by (1) factors breaking dormancy, (2) factors inducing dormancy, (3) factors coinciding with conditions required for germination, (4) factors different from conditions required for germination and/or inhibitory for germination. (From Karssen, C. M., *The Physiology and Biochemistry of Seed Development, Dormancy and Germination*, Khan, A. A., Ed., Elsevier, New York, 1982, 243. With permission.)

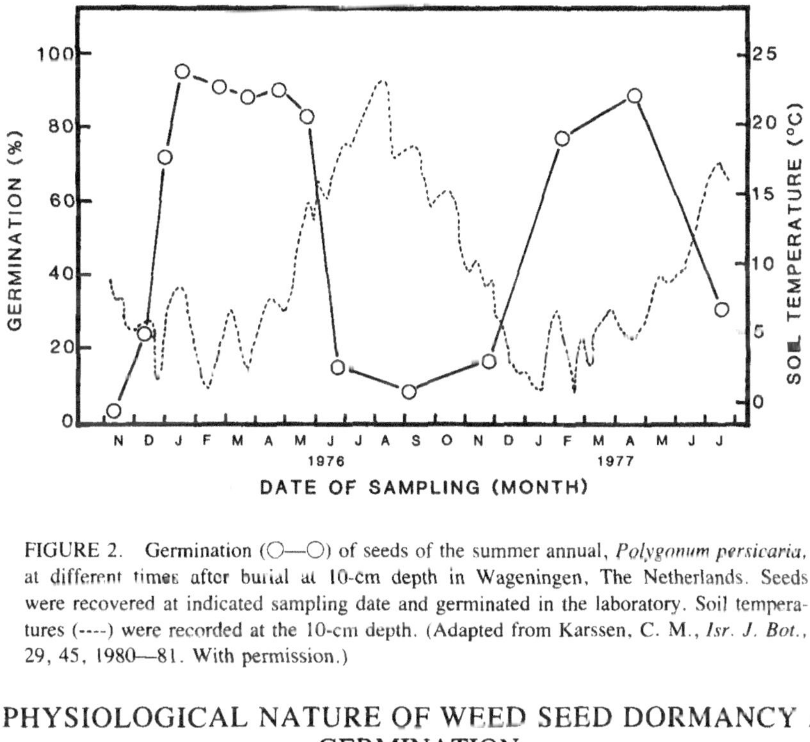

FIGURE 2. Germination (O—O) of seeds of the summer annual, *Polygonum persicaria*, at different times after burial at 10-cm depth in Wageningen, The Netherlands. Seeds were recovered at indicated sampling date and germinated in the laboratory. Soil temperatures (----) were recorded at the 10-cm depth. (Adapted from Karssen, C. M., *Isr. J. Bot.*, 29, 45, 1980—81. With permission.)

II. PHYSIOLOGICAL NATURE OF WEED SEED DORMANCY AND GERMINATION

A. Coat-Imposed Dormancy

In many weed seeds, dormancy is imposed by coats or coverings over the embryo.[13-15] These coverings (seed coats, fruit coats, endosperm) may prevent germination by: (1) blocking water entry to the embryo; (2) limiting exchange of respiratory gases; (3) blocking light penetration to the embryo; (4) mechanically constraining embryo expansion; and/or (5) preventing loss of inhibitors from the embryo. One or more of these conditions may occur in a particular seed.[10]

1. Seed Coat Impermeability to Water

The prevention of water uptake into the seed that has an impermeable seed coat is a very effective dormancy mechanism occurring in weedy members of several families, particularly those of the Leguminoseae, Malvaceae, Chenopodiaceae, and Solanaceae.[13] Many of the most troublesome and long-lived weed seeds have coats that are impermeable to water.[4,5] These seeds, also termed hard seeds, respond to scarification of the seed coat by imbibing water and germinating.

The physiological processes by which weed seed coats become impermeable and then later become permeable are not completely understood, but it is evident that the coats become impermeable during the later stages of development on the parent plant.[16-18] Marbach and Mayer[17] reported that the development of seed coat impermeability of a wild pea (*Pisum elatius*) strain correlated with seed coat browning in the presence of atmospheric O_2. They concluded that polyphenol oxidase catalyzed the polymerization of water-impervious polymers in the seed coat. Recent research with *Sida spinosa* indicates that peroxidase-catalyzed formation of lignin from phenolic compounds in the palisade layer of the developing seed coat correlated with the onset of seed coat impermeability.[16] Immunocytochemical methods were used to demonstrate the presence of peroxidase and the absence of polyphenol oxidase in *S. spinosa* seed coats. Thus, peroxidase and not polyphenol oxidase was implicated in the formation of water-impervious polymers in *S. spinosa* seed coats (Figures 3 and 4).[16] As the developing seed of *S. spinosa* dehydrated to about 10% moisture, the chalazal slit sealed and completed the impervious barrier to water.[16] Hyde[19] earlier reported that water loss during final dehydration of some maturing seeds of Papilionaceae was controlled by the water status of the cells surrounding the hilum. Peroxidase and not polyphenol oxidase is the key enzyme involved in seed coat lignification of the seeds of two other malvaceous species (*Anoda cristata* and *Abutilon theophrasti*) and of two leguminous species (*Sesbania exaltata* and *Crotalaria spectabilis*).[20] In each instance, lignification of the seed coat correlated well with the onset of seed coat impermeability to water.

The change from coat impermeability to permeability does not involve the entire coat and does not occur at random over the seed coat surface, but involves specific areas of the seed coat. Ballard[15] pointed out that the strophiole of some legumes contains an area of weakness that separates and permits water to enter the seed. The heating of hard seeds of the legume *Albizia lophantha* causes the expulsion of a strophiolar plug from the seed coat (Figure 5).[21] The plug is an area of the palisade layer that is shed before the seed imbibes water and germinates. Water enters *Abutilon theophrasti*[22] and *Sida spinosa*[23] seeds through the chalazal area (Figure 6).

In *S. spinosa,* the separation of the portion of the seed coat from the underlying layer of subpalisades in the chalazal region always precedes water uptake. This portion, also termed a blister because of its appearance when formed on dry seeds, separates when the side walls of the subpalisades break at thin areas near the top of the cells (Figure 7). These thin sections of the walls are due to a lack of secondary wall development and appear to be weak sites that enable the blister to form and trigger water entry into the embryo of previously hard seed.[24] That the subpalisade cell layer occurs only in the region where the blister forms, and that the subpalisade cell walls are specialized for breakage, indicate that the subpalisades are instrumental in the change from coat impermeability to permeability.[23,24] Further evidence indicates that the palisade layer of the blister region of the *S. spinosa* seeds is less intensely lignified than palisades of the seed coat at other areas, and that cellulose components of the palisades in the blister area are more readily degraded by cellulase than are the palisades in other coat regions.[25] Thus, the palisade-subpalisades of the chalazal region of *S. spinosa* seeds have properties that facilitate the acquisition of permeability.

Alternate cycles of temperature or hydration have been suggested as natural means of increasing permeability of seed coats to water.[26] Stresses, imposed by contraction and

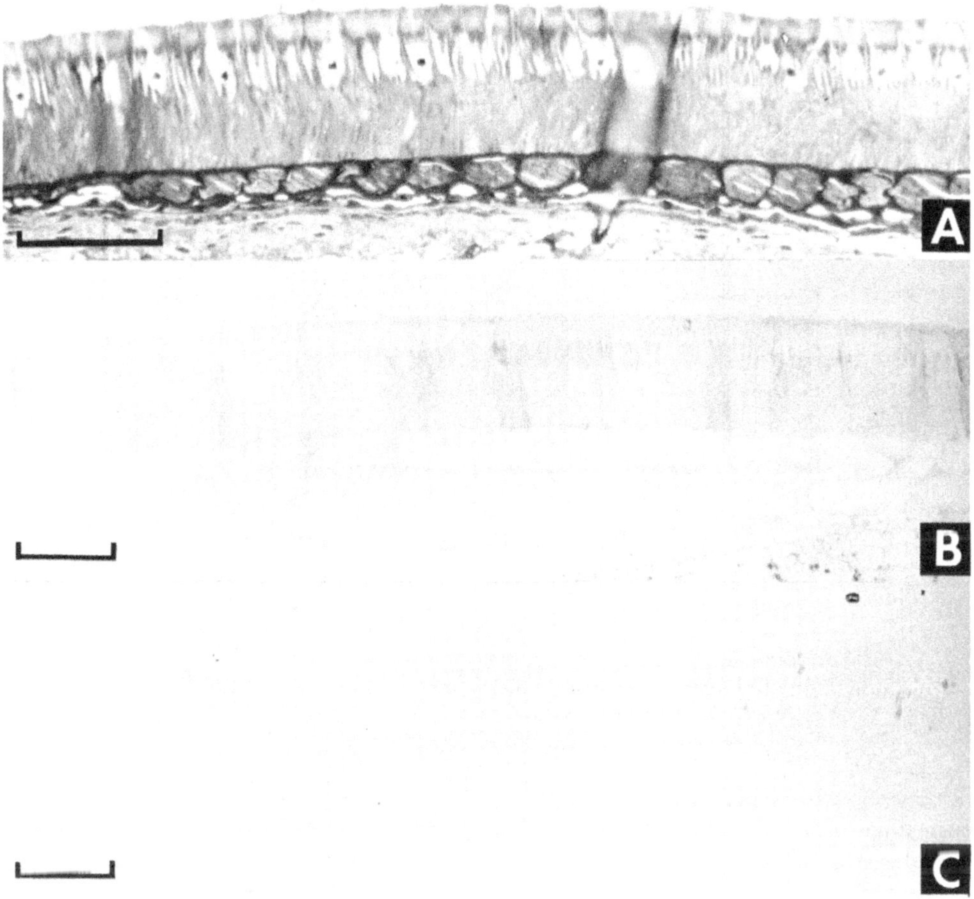

FIGURE 3. (A) Portion of *Sida spinosa* seed (12 days post anthesis) stained wih toluidine blue to show palisade (P) layer and underlying tissues; (B) portion of *S. spinosa* seed (16 days post anthesis) stained for peroxidase using the antibody technique of Sternberger. Note positive reaction for peroxidase in palisades; (C) portion of *S. spinosa* seed (16 days post anthesis) stained for polyphenol oxidase using antibody technique of Sternberger. Note negative reaction. Scale = 50 μm. (From Egley, G. H., Paul, R. N., Jr., Vaughn, K. C., and Duke, S. O., *Planta*, 157, 224, 1983. With permission.)

expansion due to temperature or moisture changes, on the weak sites in impermeable coats may trigger cell separations and/or ruptures, enabling water entry. Afterripening at constant, warm temperatures under laboratory conditions also increases seed coat permeability to water, perhaps through increased fragility of cell walls at weak sites in aged seeds.[24]

The location of barriers to water penetration through the seed coat has not been precisely determined.[13,14] The waxy coatings and the cuticles over the surface of seeds, as well as cuticles within seed coats, have been considered, but there is little evidence to establish them as universal barriers to water uptake.[15] In some legumes, the inner (proximal) portion of the palisades was suggested as the barrier.[15] However, the barrier in seeds of various species needs to be identified since it may differ among species. Regardless, it is becoming evident that the loss of seed coat impermeability to water does not occur generally over the seed but occurs in specific, specialized coat regions.[15,21]

2. Seed Coat Barriers to Gas Exchange

The intact seed coat may also reduce O_2 diffusion to the embryo. Seed coat scarification

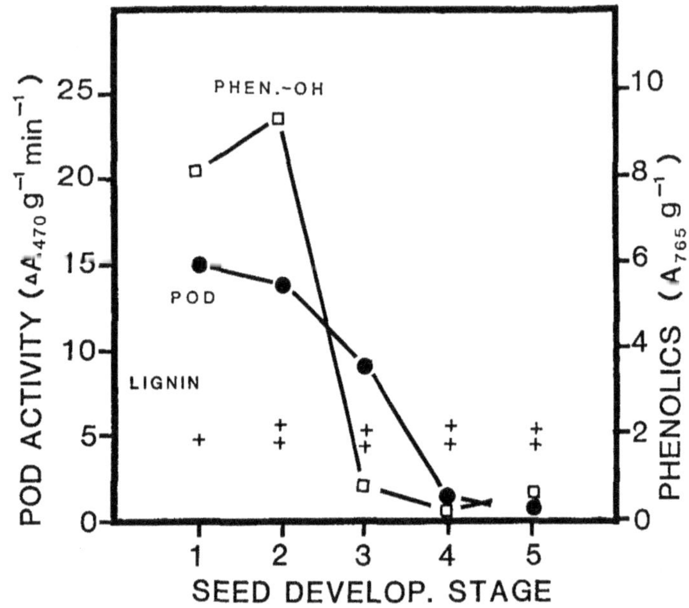

FIGURE 4. Peroxidase (POD) activity and hydroxyphenolic levels in extracts from seed coats of *(Sida spinosa)* at five stages of seed development. Lignin presence indicated by positive test with phloroglucinol. Developmental stage (days post anthesis) 1, (10 to 12); 2, (13 to 14); 3, (15 to 16); 4, (17 to 18); 5, 20 to 25). (From Egley, G. H., Paul, R. N., Jr., Vaughn, K. C., and Duke, S. O., *Planta*, 157, 224, 1983. With permission.)

or removal may expose the embryo to the increased O_2 levels necessary for germination.[10] Sometimes incubation of intact seeds in high O_2 tensions increases germination, suggesting that oxygen levels within the seeds are insufficient for germination (see Bewley and Black[10] for discussion of coat-imposed dormancy). Additionally, seed coat barriers to O_2 uptake might also prevent oxidative destruction of germination inhibitors within the seed. The O_2 requirements for breaking dormancy might differ, depending upon the inhibitor level or respiratory requirements within the seed. Porter and Wareing[27] concluded that O_2 diffused equally well through the coats of the two dimorphic *Xanthium pensylvanicum* seeds of different dormancy levels. Thus, the smaller, dormant seed of the two-seeded fruit was not dormant because of the differential restriction to O_2 diffusion through the seed coat. However, the O_2 requirement of the smaller dormant seed was greater and it, therefore, was more responsive than the nondormant seed to increased external O_2 levels.[27]

Water in the imbibed seed coat may itself be a barrier to O_2 diffusion.[10] Although water may not be completely blocking O_2 uptake, the limited solubility of O_2 in water (2.8% at 25°C) might lower the O_2 levels reaching the embryo below the thresholds needed for germination.

Mucilagenous seed coats may also restrict O_2 diffusion to the embryo and internal seed tissues.[13,28] Phenolics in coats may combine with O_2 diffusing through the coat, thus, reducing the O_2 levels reaching the embryo.[29,30] The possibility that O_2 reaching the embryo might be limited through oxidation of phenolics in the coat is interesting in that it could constitute a timing mechanism that delays germination until phenolics in the coat are reduced below a critical level or until the oxidase enzyme activity declined. However, there are instances where weed seeds germinate under near anaerobic conditions, which indicate that O_2 effects in germination are complex (see Section II.C.2).

Seed coat restriction to outward diffusion of CO_2 from the respiring embryo and the build-

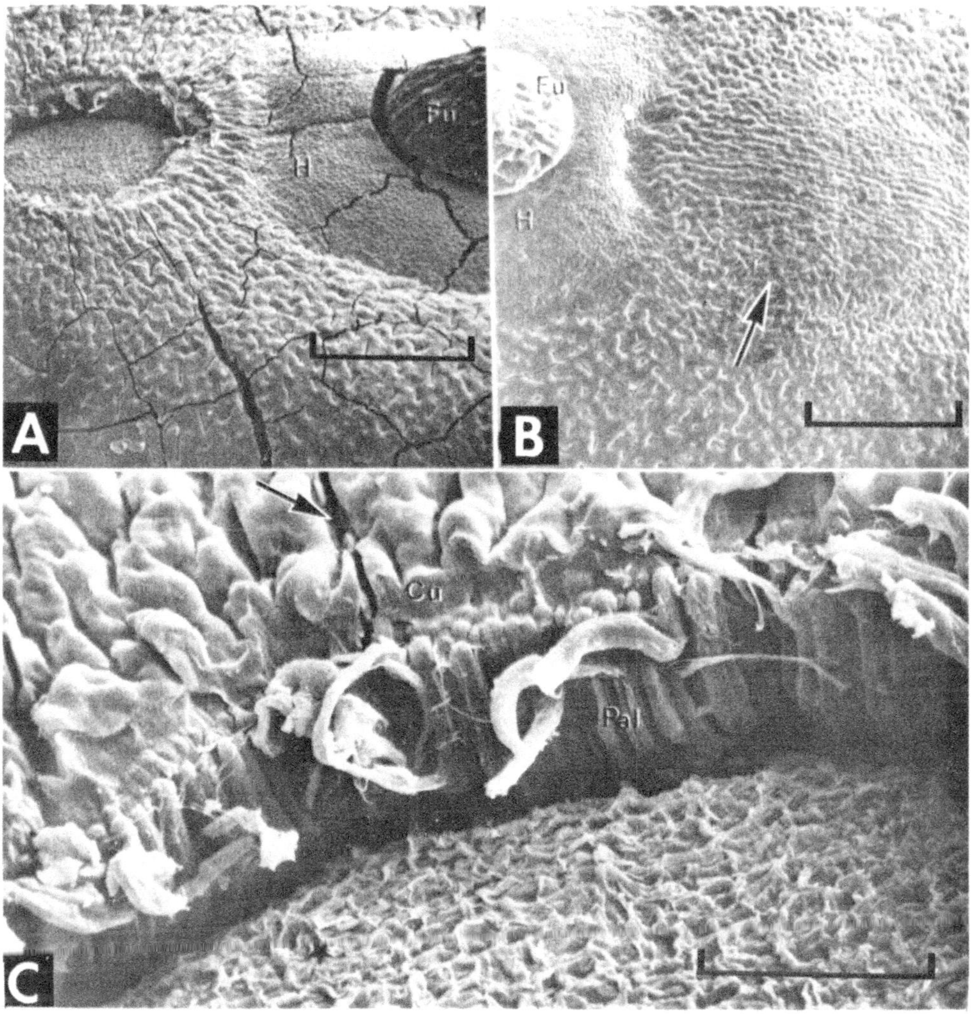

FIGURE 5. (A) Strophiolar region of *Albizia* seed. Heat-treated seed showing cavity where the strophiolar plug has erupted. The cavity is adjacent to the funiculus (Fu) and hilar tissue (H). Scale = 250 μm; (B) strophiolar region prior to eruption of plug (arrow). Scale = 250 μm; (C) detail of cavity. The palisade cells (Pal) are overlain by a thick cuticular layer (Cu) which is deeply folded (arrow). Broken, thin cell walls are evident on the floor of the cavity. Scale = 50 μm. (From Dell, B., *Am. J. Bot.*, 67, 556, 1980. With permission.)

up of inhibitory CO_2 concentrations within the seed was considered earlier as a possible means of inhibiting germination and maintaining dormancy. However, evidence in support of this role of CO_2 in dormancy is generally lacking.[10]

3. Mechanical Constraints to Embryo Growth

Cuts or punctures through seed coats over or near the radicle are often more effective in stimulating seed germination than coat ruptures elsewhere on the dormant seed.[31-34] These results can be explained as due to the release of a mechanical constraint to radicle elongation.[31,32]

Typical germination (radicle emergence from the radicle end of the seed) of *Phacelia tanacetifolia* and *Striga asiatica* seeds occurs when the coats are pierced over the radicle, but atypical germination (radicle emergence from the cut end, not from the radicle end of the seed) occurs in a high percentage of the seeds when the coats were cut or pierced elsewhere.[31,32] In each instance, scarification induces germination but the radicles are unable

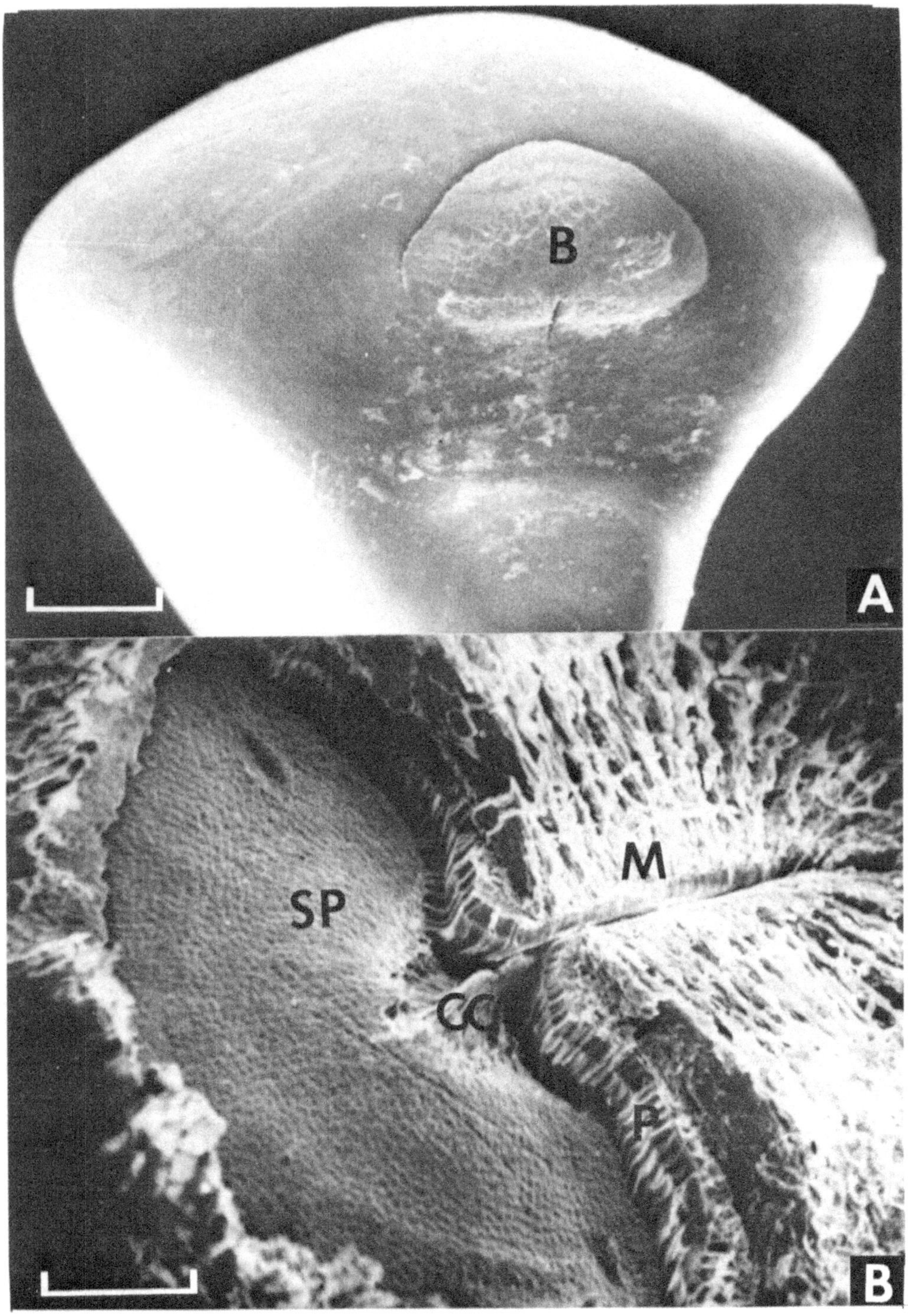

FIGURE 6. (A) Scanning electron micrograph (SEM) of mature *Sida spinosa* seed showing "blister" (B) that formed near chalazal slit. Scale = 250 μm; (B) SEM of *S. spinosa* seed with "blister" removed, exposing subpalisade (SP) layer with chalazal cap (CC) cells in chalazal slit, and palisade layer (P) outside area defined by the blister. Scale = 100 μm. (From Egley, G. H. and Paul, R. N., Jr., *Am. J. Bot.*, 68, 1056, 1981. With permission.)

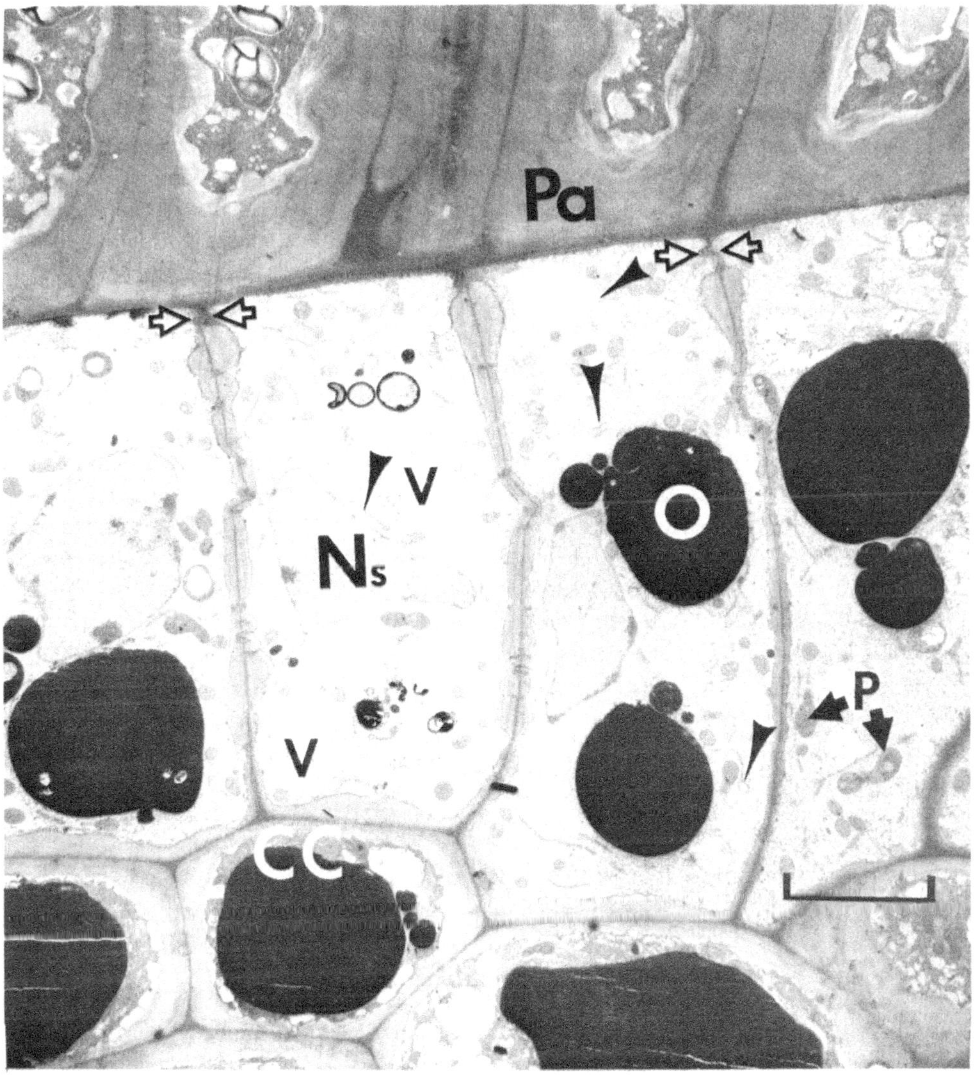

FIGURE 7. Transmission electron micrograph of subpalisade region of *Sida spinosa* seed at 6 days post anthesis. Hollow arrows point to thin cell wall portions where future breakage occurs prior to imbibition of water. The subpalisade layer is between palisades (Pa) and chalazal (CC) cells. OV, osmiophilic vacuole; V, vacuole; Ns, subpalisade cell nucleus; P, plastids. Scale = 5 μm. (After Egley, G. H. and Paul, R. N., Jr., *Am. J. Bot.*, 69, 1402, 1982. With permission.)

to protrude through the intact covers over the radicle. Instead, either the radicle or the cotyledon end of the embryo emerges from the cut portion of the seed. In the case of nondormant germinating *Lactuca sativa* seeds, radicle protrusion is prevented if the endosperm over the embryo remains intact.[35] It was proposed that softening of the coverings over the radicle by hydrolases normally permits radicle protrusion.[32] However, cell wall degradation of *L. sativa* endosperm, which is considered to be the restraining tissue in the seed, was not detected prior to radicle protrusion.[36] Thus, it is unlikely that cell wall degradation of the constraining endosperm is a prerequisite for radicle protrusion. On the other hand, the restraining tissues (aleuron layer) over the radicle of *S. asiatica* seeds are measurably weakened about 2 h before radicle protrusion.[31] In the latter case, weakening of the aleuron is a prerequisite for germination, although the anatomical and physiological bases for the weakening were not determined.

Psaras et al.[37] showed that, although endosperm walls of *L. sativa* remained unchanged, the endosperm cells over the radicle tip undergo structural changes (cells swell, become vacuolated, and reserves are degraded) in the cell just prior to radicle protrusion. This change occurs only over the radicle tip in response to red light (seeds destined to germinate) and do not occur after a red to far-red light treatment (seeds remain dormant). It was not determined if structural changes are responsible for the weakening of the endosperm, but the changes indicate preparation of the endosperm in advance of radicle protrusion and suggest that the endosperm cells are reacting to a germination signal. Possibly the changes are the early stages of endosperm degradation of reserve foods necessary for the germinating embryo. Conceivably, massive cell wall degradation is not necessary to weaken the endosperm layers if internal cellular components are degraded and the water potentials altered so that rapid influx of water, cellular swelling, and disruption may occur.

Current evidence permits some speculation as to the possible pregermination signals which could trigger changes in the endosperm or seed coat layers. Bewley and Halmer[38] suggested that hormones originating in the embryos of *L. sativa* controlled mannanase production in the endosperm which, in turn, degraded the major constituents of the endosperm cell walls. Although these were actually post-germination events, it is possible that the first subtle steps in the degradation process may result in radicle protrusion. A low level of ethylene during the first 12 h of *L. sativa* seed incubation is produced just prior to radicle protrusion and is essential for germination.[39] *Avena fatua* seeds also produced a low level of ethylene prior to germination.[40] Dunlap and Morgan[39] suggested that this pregermination production of ethylene might be involved in the preparation of the *L. sativa* endosperm for radicle protrusion. This is conceivable because ethylene promoted the release of the hydrolase from endosperm cell walls of *Hordeum vulgare* seeds.[41] That ethylene could trigger these observed endosperm changes is further indicated by evidence that stress imposed by physical impedence of expanding cells stimulates ethylene evolution.[42] Thus, the physical restraint imposed upon the expanding embryo may itself cause the production of ethylene, which could serve as the signal to initiate enzymatic activities that cause structural changes in cells of the endosperm layer. Still, the causal factor of the expansion of the embryo to the point of pressing against the endosperm would remain unexplained. This view suggests that several signals may be involved in the germination process.

It is evident that consideration of the mechanical constraint theory then becomes one of balancing the ability of the radicle to grow against the constraint imposed by the coverings over the radicle. Esashi and Leopold[43] showed that the resistances imposed upon the radicle by the coats of both dormant and nondormant cocklebur seeds were similar, but the growth force attained by the embryos of the two seeds were dissimilar with the dormant embryo developing the lesser growth force. Despite this, the mechanical constraint theory has gained creditability.

Quite obviously, seed coats and other structures surrounding the embryo are extremely important to both the survival and the regulation of germination of weed seeds. Because the seed coat is the embryo's major line of defense against the hazards of the environment, it is not surprising that it should also impart barriers to early and rapid germination. Thus, the covering structures around the embryo may serve a dual function: (1) protection of the embryo and (2) regulation of the time of seed germination.

B. Endogenous Regulation of Weed Seed Dormancy and Germination

1. Chemical Regulation

a. Hormones

The hormonal regulation of seed dormancy and germination has foundation in the inhibitor-promotor theory of plant growth and development regulation.[44] Much of the evidence to support application of this theory to weed seeds is based on the stimulation or inhibition of

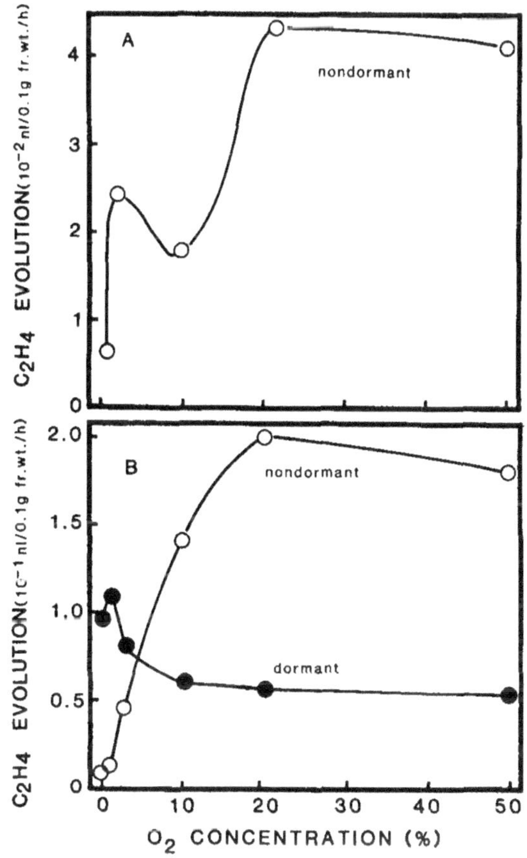

FIGURE 8. (A) C_2H_4 production of nondormant axial segments of *Xanthium pensylvanicum* seeds in response to various O_2 concentrations at the beginning of the 9th hr of incubation; (B) C_2H_4 production of nondormant and dormant axial segments of *X. pensylvanicum* seeds in response to various O_2 concentrations at the beginning of the 15th hr of incubation. (After Esashi, Y., Ohhara, Y., Kotaki, K., and Watanabe, K., *Planta*, 129, 23, 1976. With permission.)

germination by applied hormones (gibberellins, ethylene, cytokinins, or abscisic acid). An important test for roles of hormones in seed dormancy and germination regulation is correlation of endogenous hormone content with the degree of dormancy or germination rates. For a hormone to be a significant regulator of dormancy, one might expect the appearance or increase of promotors or the disappearance or reduction of inhibitors prior to germination or loss of dormancy. There are few instances, however, where these changes in hormones have been correlated with the loss of dormancy or the stimulation of weed seed germination.[10]

Ethylene stimulates the germination of several weed seeds.[7,45,46] Earlier work with non-weedy species established that germinating *Ricinus communis* seeds produced ethylene.[47] Imbibed, dormant *Arachis hypogaea* seeds also produced ethylene but the internal amounts (0.4 to 0.9 $\mu\ell/\ell$) were below the threshold (1.0 $\mu\ell/\ell$) necessary for germination.[48] During incubation just prior to germination, *Avena fatua* seeds also produce a subthreshold level of ethylene that may contribute to the stimulation of germination.[40] Embryonic axes from nondormant *Xanthium pensylvanicum* seeds were capable of producing more ethylene than were axes from dormant seeds.[49] Nondormant *X. pensylvanicum* embryos produced ethylene with two ethylene-producing systems, one operating aerobically and the other anaerobically, but the dormant embryo lacked the ability to produce ethylene aerobically (Figure 8).

Removal of ethylene from seeds, either by trapping ethylene from the air around the seeds[45] or by evacuation under reduced pressure,[50] can reduce germination, providing further support for the concept of a regulatory role of ethylene in germination.

The promotion of seed germination in several species by ethylene and the production of ethylene by some seeds prior to germination indicate that the gas may play an important role in the stimulation of germination of weed seeds. It also appears that some nondormant seeds are capable of producing more ethylene than dormant seeds and that the failure of the dormant seed to germinate is due to the lack of ethylene production at a crucial stage of seed preparation for germination. Thus, the breaking of dormancy might be due to the development of the ethylene-producing system in the dormant seed. However, more evidence is necessary to support this hypothesis.

Although there are many reports that gibberellins induce seed germination,[51-53] evidence is needed to establish an endogenous role in dormancy and germination regulation.[10,52,54] Black and Naylor[55] reported that gibberellic acid (GA_3) treatment of the *A. fatua* mother plant during seed maturation prevents onset of dormancy of seeds, apparently by artificially increasing the GA_3 levels within the seed. They concluded that the low levels of GA within the seed are the basis for seed dormancy in *A. fatua*. Perhaps the most convincing evidence for participation of endogenous gibberellins in germination regulation has been obtained with some woody species.[10] *Corylus avellana* seeds, which undergo chilling to break dormancy, are able to produce more GA later when the seeds are transferred to a warmer germination temperature.[56] There is evidence that gibberellins from the embryo play a role in reserve food degradation in the endosperm[57] or cotyledons,[58] but a mode of action in breaking dormancy or inducing germination remains unclear.

Also, increases in cytokinins in dormant tree seeds[59] occur during dormancy-breaking chilling periods and after light treatment of dormant *Rumex obtusifolius*[60] and *Spergula arvensis*[61] seeds. However, the significance of the increases in extractable cytokinin to the stimulation of germination is not clear.[10,62] Applications of abscisic acid (ABA) are well known to inhibit seed germination.[54,63] However, evidence for the role of endogenous ABA in the regulation of seed dormancy and germination is also inconclusive. Gibberellins, cytokinins, and ABA occur in seeds (particularly during development) and their levels may change during maturation and germination. However, these changes do not always correlate with breaking of dormancy or stimulation of germination, and evidence to substantiate a regulatory role for hormonal control in these processes is still lacking.

Trewavas[64] emphasized that the sensitivity of plant tissues to hormones is the critical factor in the hormonal control of plant growth and development. The presence of a promotor or inhibitor has little significance if the tissues are insensitive and unable to respond. Increased sensitivity of seed tissues to hormones may be necessary for an induction process and the absolute hormone concentrations may be insignificant. Changes in sensitivity of *Amaranthus retroflexus*[65,66] and *Xanthium pensylvanicum*[67] seeds to ethylene were observed during incubation.

Another factor to consider is that the critical site of hormonal action in seeds may reside in only a relatively few cells and whole seed or whole organ extracts may not reveal subtle quantitative hormonal changes. The use of histochemical techniques may be useful for the detection of hormones in tissues, to locate hormone level shifts that may occur at target sites in the growth axis, and to correlate such effects with loss of dormancy or induction of germination. Although inhibitors or promotors are present in seeds, they may be significant if only located at target sites involved in regulating dormancy or germination.

b. Nonhormones

The inhibition of germination by compounds other than ABA (mainly phenolics and unsaturated lactones), either residing in the embryo or in tissues surrounding the embryo, has been considered.[10] Recently, endogenous shortchain fatty acids (C_6 to C_{10}) were proposed as germination inhibitors in *Avena fatua* seeds.[68] This role has not been substantiated.[69] The possible role of fatty acids in the regulation of dormancy and germination is particularly interesting because fatty acids are components of membranes that have been considered to be important in the regulation of seed germination.[51,70] Wesson and Wareing[71] concluded that volatile inhibitors were produced from weed seeds while they were buried in the soil; these inhibitors, in turn, induced a light requirement in the seeds. They stressed that CO_2 was not a volatile inhibitor rising from the seeds. Low O_2 levels in soil may be responsible for the production of acetaldehyde and other products of anaerobic metabolism that may inhibit germination and induce a light requirement in buried seeds.[72] Taylorson detected volatile compounds arising from weed seeds during accelerated afterripening, but he found no relationship of the volatiles with dormancy changes.[73]

Wareing and Foda[74] postulated that the dormant embryos from small *Xanthium* seeds contain unknown inhibitors that cannot leach through the impermeable seed coat and that washing decoated embryos removed the inhibitor and permitted germination. Incubation of intact seeds in high O_2 levels resulted in the oxidative destruction of the inhibitor. Those results gave earlier support for the inhibitor hypothesis of dormancy control in *Xanthium;* however, recent results with anaerobic effects on ethylene production have raised some doubt about the role of inhibitors in the coat-imposed dormancy.[75]

Leachates from *Striga asiatica* seeds contained an unidentified germination inhibitor.[76] Also, numerous inhibitors such as phenolic acids and shortchain peptides have been detected in leachates from a number of weed seeds but their role in self-regulation of seed dormancy is unknown. The allelopathic effects of these leachates on other seeds are considered in Section II.C.6 and in Chapter 5.

2. *Metabolic Control of Germination and Dormancy*

Metabolic processes are dependent on water, and since dry seeds contain only about 10 to 15% water, metabolic processes in dry seeds are greatly attenuated. The extent of synthetic or other metabolic maintenance activities in dry seeds is not well understood because of the difficulties in studying such processes in dry tissues.[77] For example, in vitro study of constituents derived from dry seeds with aqueous extraction techniques may invalidate any extrapolations of results to in vivo activities. Use of ^{14}C-labeled ethanol in the gaseous phase has been used to a limited extent to probe metabolism in dry seeds of dry *A. fatua* seeds.[77] This technique may result in artifacts, because ethanol is known to have effects on membranes and dormancy.

The ultrastructure of embryonic cells of dry seeds is relatively simple, with underdeveloped mitochondrial cristae and little endoplasmic reticulum.[78] The cells of dry seeds of some species lack mitochondrial electron transport components. Very few or no ribosomes are found in polysomes of dry seeds. Slow changes in dry seed constituents, such as lipids, may be due to either slow metabolic processes or to nonenzymatic chemistry.

More is known about the metabolic activities of imbibed, dormant seeds. In *Agrostemma githago,* DNA synthesis is very low in dormant seeds.[79] During afterripening, there is an increase in the potential for DNA synthesis in imbibed seeds, which precedes loss of dormancy. During early imbibition, dormant and afterripened seeds of *A. githago* synthesize approximately equal amounts of protein and RNA which are qualitatively similar.[80] This, and the finding that there is little difference between the methylation of RNA, as well as tRNA and rRNA synthesis, in dormant and afterripened *A. githago* embryos, indicate that dormancy is related to a specific, rather than to a general, block of protein synthesis.[81] Differences between the proteins synthesized during imbibition of dormant and afterripened *A. githago* embryos begin to occur about the time that DNA synthesis starts to increase in afterripened embryos.[82] After imbibition, protein synthesis by the embryos of dormant *A.*

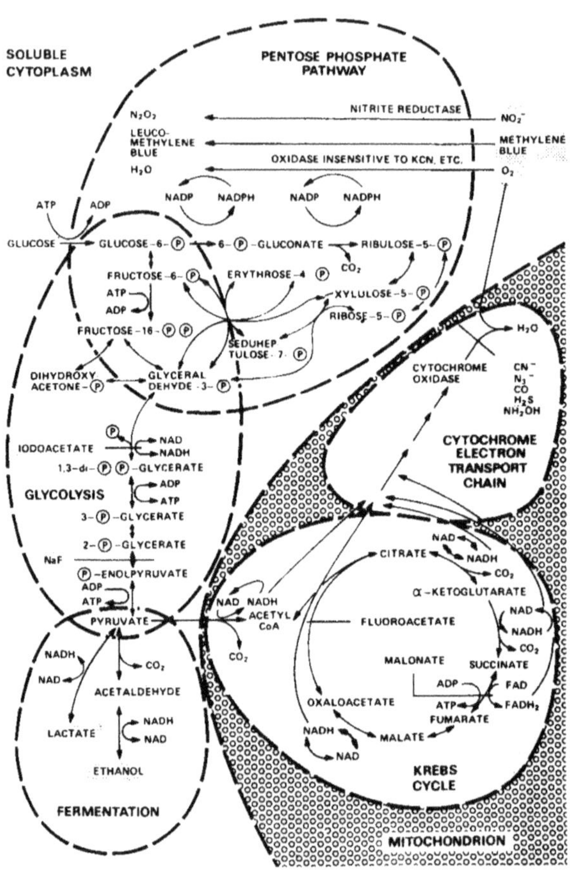

FIGURE 9. Respiratory metabolic pathways. Substances in boxes stimulate germination in dormant seeds. The diagram illustrates how such substances could directly or indirectly stimulate the activity of the pentose phosphate pathway. (From Roberts, E. H. and Smith, R. D., *The Physiology and Biochemistry of Seed Dormancy and Germination*, Khan, A. A., Ed., Elsevier/North Holland, New York, 1977, 385. With permission.)

githago seeds decreases with time, whereas embryonic protein synthesis of nondormant seeds accelerates. On a whole seed basis, however, the rates of RNA and protein synthesis of dormant and afterripened seeds are similar for 24 to 48 h after beginning imbibition at temperatures favorable for germination.[82,83]

Alteration of the respiratory pathways of imbibed, dormant seeds is concomitant with loss of dormancy (Figure 9).[84-87] Shifts of respiration from the Krebs cycle and the mitochondrial electron transport chain to the pentose phosphate pathway (PP pathway) or from the mitochondrial cyanide-sensitive pathway to the alternative cyanide-resistant respiration pathway have been implicated in the loss of dormancy in seeds of several species of weeds and crops.

Evidence for involvement of a switch from Krebs cycle/electron transport respiration to the PP pathway in breaking of dormancy is indirect. Inhibitors of mitochondrial electron transport, the Krebs cycle, and of glycolysis have been demonstrated to remove dormancy. One possible effect of these inhibitors is to increase O_2 consumption by the PP pathway and to decrease it by the electron transport system of the mitochondrion. In some cases, a shift to the PP pathway by dormancy-breaking inhibitors has been indicated by a reduced C_6/C_1 ratio[84] — an indication of more carbon moving through the PP pathway than through the glycolytic pathway. It has been hypothesized that oxidation of NADPH may limit the rate of PP pathway activity in dormant seeds, because compounds that utilize NADPH in their reduction, such as nitrate, nitrite, and methylene blue, induce germination.

The role of the PP pathway in the breaking of dormancy has recently been questioned because activities of PP pathway enzymes do not change until germination has already begun,[88] and C_6/C_1 ratio changes do not always fit the theory.[89,90] As discussed by Bewley and Black,[10] however, the validity of using C_6/C_1 ratios to estimate amounts of glucose entering glycolysis or the PP pathway is controversial. Nevertheless, new evidence with *Avena fatua* seeds indicates that breaking of dormancy with inhibitors of mitochondrial respiration is due to stimulation of the alternative respiration pathway, rather than a shift to the PP pathway.[91]

Esashi and colleagues have developed considerable data to support the view that an appropriate balance between the cyanide-sensitive cytochrome and the cyanide-resistant alternative pathways must be present for germination of *Xanthium* seeds.[87,92-94] Inhibitors of the alternative respiratory pathway such as benzohydroxamic acid and salicylhydroxamic acid inhibit germination, whereas inhibitors of mitochondrial respiration stimulate germination. Stimulation of germination of *Xanthium* seeds by ethylene, O_2, and alternating temperatures has been linked to increased alternative respiratory pathway activity. One means by which increased respiration by the alternative pathway may stimulate germination is through enhanced production of adenylate pools, allowing for increased ATP levels via the mitochondrial cytochrome system.[94]

C. Effects of the Natural Environment on Weed Seed Dormancy and Germination

1. Light

Light has long been known to be a requirement for the germination of the seeds of many weed species.[95,96] Seeds of most weed species are small and contain relatively little food reserve, so germination on or near the soil surface is necessary for seedling survival. The light requirement is a principal means by which seed germination is restricted to the proximity to the soil surface. Most soils attenuate light very effectively, so that sufficient light to induce germination is present only in the first few millimeters from the soil surface.[97,98] Thus, in an agricultural environment, many weed seeds are buried by cultivation and can germinate only when reexposed to light by subsequent cultivation. The increased emergence of weed seedlings caused by soil disturbances is a well-known phenomenon. The effectiveness of light in causing germination after exposure by cultivation is strongly dependent on the season, however.[99]

Phytochrome, the ubiquitous pigment controlling photomorphogenesis in higher plants, is thought to control most, if not all, responses of seeds to light.[100] This photochromic (i.e., photoreversible) pigment generally initiates germination when in the far-red light (around 730 nm) absorbing form and either prevents or has no effect on germination when in the red light (around 660 nm) absorbing form.[101] The photochromicity is demonstrated by giving brief (usually 10 min or less) alternating sequences of phytochrome-saturating red and far-red light. Germination is dependent only on the last light treatment.

In an agricultural environment the seed is usually exposed to light of rather complicated spectral composition for prolonged periods, so laboratory studies may not accurately reflect responses of seeds to light in a natural environment. Shading by vegetation, for instance, results in light with a relatively low-red to far-red photon flux ratio, producing relatively low P_{fr} to P_r ratios in underlying seeds.[98,102-106] In an extensive study, Gorski et al.[107] found that light filtered through leaf canopies of various species inhibited or retarded germination of seeds of many species that readily germinated under diffuse white light.

The spectral characteristics of the seed coat can also influence the P_{fr}/P_r ratio produced in the seed by sunlight. For instance, *Amaranthus* spp. seed coats have been found to transmit

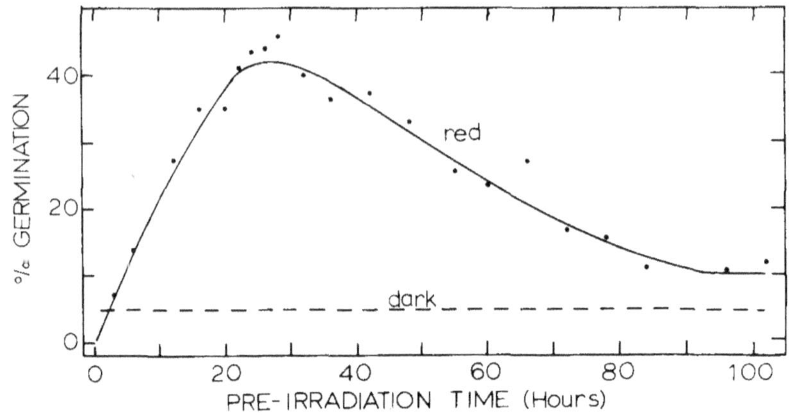

FIGURE 10. Changing germination response with dark incubation time of *Portulaca oleracea* seed to 1 h of red light. (From Duke, S. O., Egley, G. H., and Reger, B. J., *Plant Physiol.*, 59, 244, 1977. With permission.)

far-red better than red light.[108] Variability in seed coat light transmission (both quantitative and qualitative) is partially responsible for variation in sensitivity to light among seeds of a population (see Chapter 1).

Considerable variation exists among weed species in the pattern of development of photoblastic properties. As in the case of dormancy, seeds of some species are initially produced as photoblastic (e.g., *Amaranthus retroflexus*),[109] others develop the property (e.g., *Rumex crispus* and *Echinochloa crus-galli*),[109] and in others it can be induced (e.g., *Bidens pilosa*).[110] The light requirement is lost with time in all cases.

In photoblastic seeds the sensitivity to light changes with time in darkness after the beginning of imbibition of dry seeds (Figure 10).[111,112] In all cases studied, sensitivity increases relatively rapidly (but not as fast as water uptake),[113] peaks, and then declines slowly, relative to the initial increase. This decline is the onset of a form of secondary dormancy.

The cause of this decline in response to light is not yet clear. Two principal schools of thought exist: (1) the decline is due to loss of phytochrome[114] and (2) it is due to a reduction of the effectiveness of P_{fr} in inducing germination.[111] Variations in the second hypothesis involve reduced gaseous exchange,[115] formation of inhibitory substances,[11] and decreased interaction between P_{fr} and 'X', the unknown primary receptor of P_{fr}.[116,117] The latter case is supported by the lengthening of the "escape" time (the time required after red light treatment for far-red light to become ineffective in preventing P_{fr} from causing germination) as light becomes less effective in causing germination.[111,117] Fluence-response kinetics of *Rumex crispus* seeds during various phases of the onset of secondary dormancy indicated that the loss of response to light was due neither to changes in phytochrome nor to X levels.[118] The loss is temperature sensitive, being positively correlated with temperature within a physiological temperature range.[116,119-121] This indicates that the mechanism for production of secondary dormancy is metabolic in nature.

The mechanism by which phytochrome induces germination is unknown, but its primary action is thought to be through direct interaction with cellular membranes.[122] The time required for phytochrome to induce germination after conversion to P_{fr} varies. However, the average time is generally several hours, as measured by "escape" from far-red light reversibility. Some effects of P_{fr} on the dormant embryo are apparently relatively rapid, occurring within minutes or seconds of P_{fr} formation.[123,124] These processes apparently do not induce germination because subsequent far-red light inhibits germination.

Considerable evidence indicates that phytochrome induces germination by enhancing the expansive force of the embryonic axis beyond the restrictive forces of the seed coat.[125,126] Removal of the seed coat usually removes the light requirement for germination. For the embryo to overcome the restrictive force of the seed coat or other surrounding structures, cell walls have to be loosened, osmoticum has to be formed, and, probably, carbohydates and other storage compounds have to be mobilized. The interrelationships among these processes and their association with P_{fr} are not clear.

Prolonged or high intensity light can often inhibit germination of otherwise photoblastic seeds.[127] This process is thought to involve inhibition by cycling phytochrome rapidly between P_r and P_{fr}. In a very few cases, brief phytochrome-transforming red light exposures have been demonstrated to inhibit germination.[128]

2. Oxygen, Carbon Dioxide, and Ethylene

Although metabolic processes involved in plant growth have high O_2 demands, growth can also occur for limited periods under anaerobic or near anaerobic conditions. The O_2 requirements for dormancy breaking and for initiation of germination are not clear. In fact, anaerobiosis or reduced O_2 levels may induce germination of some seeds, including those of several weed species.[75,129,130] Seeds of *Viola* sp. and *Veronica hederaefolia* were induced to germinate under N_2 or 2% O_2 but not under 8% O_2.[131] Out of 24 families, 43 genera had seeds that germinated in water.[132] Seeds of *Cynodon dactylon* and *Trifolium repens* germinated under O_2-depleted water, indicating the ability to germinate under near anaerobic conditions.[132] On the other hand, O_2 was necessary during chilling for the dormancy release of *Ambrosia artemisiifolia*.[133] Oxygen was also required for the breaking of seed dormancy in *Xanthium pensylvanicum*[134] and for the warm temperature, dry afterripening of *Avena fatua* and some other cereal seeds.[135] Therefore, oxidative reactions of some type are essential for the release of dormancy during chilling or during afterripening of at least some weed seeds. The promotive effect of elevated O_2 levels on intact seeds or of exposure of naked embryos to atmospheric levels of O_2 may be due to the impermeability of the intact coats to O_2 or due to the utilization (e.g., oxidation of phenolics) of O_2 by the coats. The possible seed coat-O_2 effects were considered in Section II.A.2.

Definitive information concerning the influences of O_2 on germination of weed seeds was collected by Esashi and co-workers from experiments with embryos of *X. pensylvanicum* seeds.[49,134] They found that both anaerobic and aerobic systems for ethylene production existed in the embryos (Figure 8). The anaerobic system operates in both types of seeds and is stimulated by 0.5% O_2 and depressed by O_2 levels above 2%. Oxygen at 0.5% also was optimal for stimulation of germination. The aerobic system was active later in germination (still prior to radicle protrusion) and was lacking in the dormant axes. The aerobic system of ethylene production was necessary for maximal germination of the nondormant *X. pensylvanicum* seeds but was not necessary if the aerobic system was fully operative. Apparently, the lack of the aerobic system is important to *X. pensylvanicum* seed dormancy.

Ballard and Lipp[129] reported that O_2 levels below 5% stimulated germination of *Trifolium subterraneum* seeds. Subsequent transfer of seeds into air was necessary for maximum germination. Anaerobiosis may have permitted production of some compound later in air which must reach threshold levels for germination. In the photoperiodically induced germination of *Eragrostis* spp. seeds, anaerobiosis during the dark period was stimulatory.[136] However, O_2 during the light period was necessary for maximum germination. It is evident that atmospheric levels of O_2 are not required for stimulation of germination in all situations. Evidently, O_2 requirements of the same seeds may differ with time according to their incubation condition and their dormancy level.

Seeds must probably tolerate a low O_2 level during the early stages of germination before seed coat rupture.[137] This is likely because even with seeds with coats that are highly permeable to O_2, the O_2 levels within fully hydrated, imbibed seeds are no greater than the limit of O_2 solubility in water. If the O_2 level is further reduced by utilization in coats or endosperm during diffusion into the seeds, it is highly possible that O_2 levels at the growth axes could be near anaerobic. A period of anaerobic respiration may occur in seeds during an early process of germination.[137-140] Respiration usually rises rapidly during the first several hours of imbibition followed by a lag period until radicle protrusion when a second rapid O_2 uptake phase occurs. This lag period may be due to anaerobiosis in the imbibed seed prior to rupture of the coat. However, not all seeds have this lag phase (e.g., *A. fatua*).[138] Lack of the lag phase may be due either to more efficient respiration or to the lack of anaerobiosis within the seed.

Kennedy and co-workers[141,142] found that *Echinochloa crus-galli* germinated well under either anaerobic or aerobic conditions. The peak rate of CO_2 evolution was delayed and the seeds were able to tolerate ethanol when germinated under anaerobic conditions. Anaerobic metabolism in *E. crus-galli* proceeds primarily through glycolysis, with partial operation of the tricarboxylic acid cycle and with little or no oxidative phosphorylation.[143] The ability of *E. crus-galli* seeds to survive under both aquatic and dry land conditions undoubtedly contributes to its seriousness as a water weed in rice. Perhaps, those seeds that are able to germinate under water or under near-anaerobic conditions are able to extend the time during which anaerobiosis can be tolerated and, thereby, reduce the energy requirements necessary to survive during periods of low energy production. This ability to tolerate anaerobiosis during germination, however, is apparently not related to the dormancy-breaking process.

Oxygen may have multiple effects, with the specific response to high or low levels depending upon the dormancy or pregermination stage. For example, high O_2 levels may destroy germination inhibitors while low O_2 levels may trigger a respiratory pathway shift that is important to germination. Further research is necessary to distinguish between these effects and to determine which occur in an individual weed seed species. Recent evidence suggests that the dormancy-breaking processes are O_2-requiring and early stages of germination may, in some instances, be stimulated by anaerobic or near-anaerobic conditions. The later stages of germination require higher levels of O_2 to provide the energy necessary for growth which occurs subsequent to radicle protrusion.

It was previously thought that CO_2 narcosis, caused by the accumulation of high levels of respiratory CO_2, was important in maintaining seed dormancy.[144] Although relatively high CO_2 levels (greater than 5%) inhibit germination of some weed seeds,[145-147] a role for CO_2 in maintaining dormancy is now regarded as unlikely. In fact, CO_2 at concentrations from atmospheric to 5% stimulates germination of seeds of some species.[148-150] Removal of CO_2 prevents germination in some instances.[151,152] The role of CO_2 in regulation of dormancy and germination is unclear but one action may involve ethylene production. Negm et al.[153] found that CO_2 and ethylene interaction are necessary to stimulate germination of lettuce seeds and that ethylene induces germination at any time if CO_2 is also present. Carbon dioxide was most effective in the stimulation of afterripened but still dormant *X. pensylvanicum* seeds in the presence of ethylene,[154] and the two stimuli interacted in the germination of *Spergula arvensis* seeds in the light.[152] However, the actions of CO_2 and ethylene were independent in the stimulation of *T. subterraneum* seed germination,[155] and CO_2 was not necessary for ethylene stimulation of *Portulaca oleracea* seed germination in the light.[156]

Seed sensitivity to CO_2 and ethylene may change with time. A CO_2-sensitive phase occurs during the first 12 h of incubation and precedes an ethylene-sensitive phase in *Spergula arvensis* seeds.[152] Carbon dioxide was also stimulatory early in *X. pensylvanicum* seed germination and preceded peak activity of the O_2-dependent, ethylene-producing system in the growth axes.[67] However, CO_2 was primarily acting late during *Amaranthus retroflexus* seed germination, while ethylene was active during all germination phases (Figure 11).[150]

Whether CO_2 plays a primary regulatory role remains to be seen, but it may enhance germination via enhanced ethylene production. There are instances, however, where CO_2

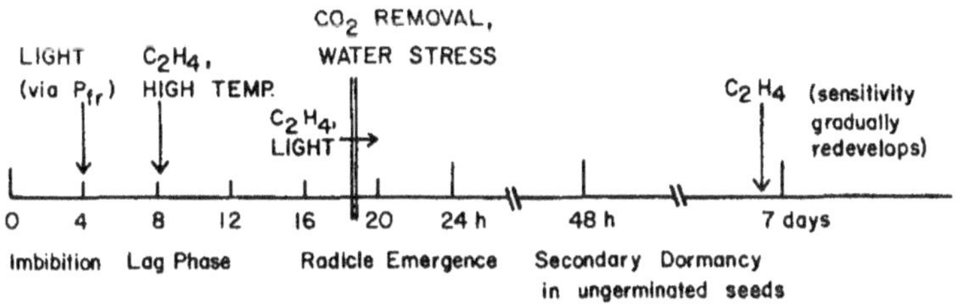

FIGURE 11. Hypothesized time sequence for *Amaranthus retroflexus* seed germination. Imbibition of water began at 0 h. Subsequent times of seed sensitivity to stimuli indicated by arrows. Time of germination indicated by double vertical line at about 19 h. Action of light occurs earliest. Ethylene is most effective just after peak sensitivity to light, but ethylene also affects radicle protrusion and release from secondary dormancy. High temperature is also most effective soon after maximum sensitivity to light. Light also relieves effects of stresses on growth just prior to radicle protrusion. Water stress affects active growth phase of germination by reducing water potential gradient between seed and environment. Lack of CO_2 also stresses active growth necessary for radicle protrusion. (From Schonbeck, M. W. and Egley, G. H., *Plant Physiol.*, 68, 175, 1981. With permission.)

is not necessary for ethylene effects. Perhaps CO_2 is not essential for ethylene production but can enhance it at certain stages of seed incubation. Carbon dioxide seems to enhance the O_2-dependent ethylene-producing system in *X. pensylvanicum*.[67] However, atmospheric CO_2 was not stimulatory to *X. pensylvanicum* seeds in secondary dormancy while application of high O_2 and ethylene levels were active.[157] This suggested that secondary dormancy of the seeds was due to the inability to produce ethylene. Burdett[158] also concluded that the decreased red light sensitivity during the pregermination period of *Lactuca sativa* seeds was associated with reduced ability to produce ethylene. Combinations of several stimuli (CO_2, ethylene, light, high temperature) were very much more stimulatory to *Amaranthus retroflexus* seed germination than any one alone or combinations of any two.[150]

Information concerning the molecular mode of CO_2 action in inducing seed germination is scant. Carbon dioxide increases ethylene production in *Oryza sativa* and *Nicotiana tabacum* leaves by directly affecting the enzyme converting 1-aminocyclopropane-1-carboxylic acid (ACC) to ethylene.[159] ACC was shown to be a precursor of ethylene in *X. pensylvanicum* seeds.[160] If CO_2 acts similarly in seeds, then it may promote ethylene which, in turn, promotes germination. The fact that CO_2 is not always promotive to germination may be due to subthreshold levels of ACC in the seed or to low sensitivity of the ethylene-producing system.

Another proposed but unsubstantiated role for CO_2 promotion of germination is dark CO_2 fixation that may produce compounds (lipids, organic acids, amino acids) necessary to support early germination.[153]

Because O_2, CO_2, and ethylene are natural components of the gaseous environment of many soils and all may influence germination of weed seeds, the effects of these gases should be given proper attention in determining the germination of seeds in nature. Although O_2 levels in soils rarely go below 19% and CO_2 levels rarely exceed 1%, it is conceivable that much greater extremes occur in microsites such as those adjacent to decaying organic matter or in flooded areas.[145] Ethylene concentrations of several ppm in soils have been recorded and this concentration could vary with differences in a microenvironment, especially at sites where ethylene-producing microorganisms flourish.[161]

3. Temperature

Germination is a complex autocatalytic process composed of many interdependent steps, most of which are enzymatic and, thus, are highly temperature dependent. Under isothermal conditions, nondormant seeds of most weed species have a single temperature optimum for germination, although bimodal temperature optima have been reported. Arrhenius plots of log of germination percent vs. 1/K are generally linear between 15 and 30°C, with a break around 30°C,[162,163] probably due to enzyme denaturation and/or lipid phase changes. In some species a break is found near 15°C,[163,164] suggesting a relationship with membrane-bound or lipid-associated enzymes.[165]

In the field, seeds rarely, if ever, experience isothermal conditions and are usually subjected to diurnal temperature fluctuations with decreasing amplitude with increasing depth in the soil. The amplitude of this diurnal fluctuation may be more important than light as an environmental signal in sensing proximity to the soil surface because it is not attenuated as effectively by soil as is light. Additionally, the temperature range is seasonally dependent. These seasonal changes in soil temperature amplitude and range apparently influence the temperature requirements for germination of weed seeds.[166] Alternating temperatures can induce germination and speed germination rate in light or dark, provided the temperature extremes are not too great.[66,167-170] This may be due to different temperature optima for different metabolic steps required for germination.[171] In five weed species, however, Taylorson and Hendricks found evidence that increases in germination caused by a brief high temperature shift are due to the action of preexisting P_{fr}.[167,170] Toole also suggested that stimulation of germination of *Agropyron smithii* seeds by alternating temperatures is due to facilitation of the proper balance and interaction of hormones, enzymes, substrates, and preexisting P_{fr}.[172] Work with *Rumex obtusifolius* seeds has shown that stimulation of dark germination by a brief 35°C treatment could be reversed by far-red light with the same fluence response curve as reversal of red-light-stimulated germination.[173] In this case, the temperature shift apparently increases sensitivity to preexisting P_{fr}.

In two *Rumex* species, the period of the higher temperature in a diurnal cycle is more critical than that of the lower temperature, germination being virtually unaffected by the lower temperature.[174] Generally, the stimulatory effect of temperature fluctuations increases with amplitude, provided that duration of the upper temperature decreases as the upper temperature increases. The time after imbibition during which a single temperature shift occurs also influences the effect of the shift, and the temperature at which the seeds are imbibed influences the time of maximum sensitivity to a single temperature shift.[169] Diurnal temperature cycles which stimulate germination are generally additive in their effects on a seed population.

Temperature strongly influences the breaking of both primary and secondary dormancy. Primary dormancy can be rapidly broken by high temperature treatment of dry seeds.[175,176] This treatment apparently greatly accelerates the afterripening process. Roberts and Smith estimated the Q_{10} for afterripening of *R. crispus* seeds to be between 3.1 and 3.9.[84]

Dormancy may also be lost in seeds stored under moist conditions at cold, but not freezing, temperatures — usually from 0 to as high as 15°C.[177] This process is referred to as stratification and is probably the principal mechanism by which weed seeds produced in the late summer and fall lose their primary dormancy in temperate climates. This process is not to be confused with prechilling, that is, stimulation of germination of afterripened seeds by a relatively short (usually not more than a few days) exposure to temperatures suboptimal for germination.[177,178] The prechilling effect has been explained by slowed thermal reversion of P_{fr} to P_r and to increased sensitivity to preexisting P_{fr}.

Both the onset and breaking of secondary dormancy are strongly affected by temperature. For instance, in *R. crispus* secondary dormancy develops more rapidly at the higher temperatures within a temperature range.[116,120] At 5°C or below, secondary dormancy does not develop and at 37°C and above the rate of induction decreases. Secondary dormancy of *R. crispus* seeds can be broken by either a prolonged cold (5°C) treatment or a brief high temperature (40°C) treatment.[170]

Nondormant seeds which germinate in darkness can often be made light requiring or forced into secondary dormancy by exposure to temperatures supraoptimal for germination while in the imbibed state. This form of secondary dormancy is termed thermodormancy[179] and is thought to result from thermal reversion of preexisting P_{fr} to P_r while in a state where P_{fr} activity is blocked.

4. Water Stress

All seeds except those with intact, water-impermeable (hard) seed coats (see Section II.A.1) imbibe or hydrate when exposed to water. Although imbibition is a prerequisite for germination, this process occurs whether the seed is nondormant, dormant, or dead. The degree and rate of imbibition influences germination and dormancy and is governed by three categories of factors: (1) properties of the seed, (2) properties of the soil, and (3) degree of contact between the seed and soil.[138]

In a steady-state situation (i.e., constant soil water potential), the final degree of imbibition results when an equilibrium between soil water potential (ψ_{soil}) and water potential of the seed (ψ_{seed}) occurs. Obviously, the ψ_{soil} determines the final ψ_{seed}, but since the mechanics of measuring ψ_{seed} and maintaining a constant ψ_{soil} over time are very difficult, only rough correlations have been measured between the two.[180] The water potentials of both soil and seed are composed of osmotic potential (ψ_o), matrix potential (ψ_m), and pressure potential (ψ_p). The soil is usually determined almost entirely by ψ_m, whereas all three factors are important to the make-up of ψ_{seed}. An important point is that ψ_{seed} is a composite of the water potentials of many cells which may differ markedly in the way they reach equilibrium with ψ_{soil}. Nevertheless, the final degree of imbibition of the seed is influenced by the complex interaction of both physical and biochemical factors as they affect each of the three components of ψ_{seed}.

After rain or watering, the ψ_{soil} in the microenvironments within which seeds exist will increase, reach a maximum, and then decline with time. Thus, the maximum ψ_{seed} that a seed reaches can be strongly dependent on the rate of movement of water from the soil to the seed. The rate of this movement is dependent on the water potential differential as well as several impedances to flow:

$$\text{Flow rate} = \frac{\psi_{soil} - \psi_{seed}}{i_{soil} + i_{seed} + i_{interface}}$$

where i_{soil}, i_{seed}, and $i_{interface}$ are the impedances within the soil, within the seed, and between the soil and seed, respectively.[138]

The rate of imbibition must be controlled in order to prevent injury. Rapid imbibition of legume seeds from which the testae (part of i_{seed}) have been removed results in pronounced damage to cell membranes, as detected by leakage of both cytosol and mitochondrial enzymes.[181] The mechanism by which i_{seed} is controlled varies greatly between weed species. For example, the testae, mucilage layers, the micropyle, and the chalazal slit are primary determinants of i_{seed} in different species.

The kinetics of initial water uptake by dry dormant or dead seeds are similar — a logarithmically declining rate of increase, followed by a plateau. In nondormant seeds this phase is followed by another increase in uptake beginning with a logarithmically increasing rate of uptake. Without sufficient initial water uptake, germination cannot occur. If the water content of the embryo does not reach the level required to support biochemical events that lead to cell expansion, germination cannot occur. The water content required for these processes is, however, lower than that required for development of adequate ψ_p (turgor) for cell expansion. There is evidence that at low ψ_{seed}, some of the physiological processes necessary for germination, such as phytochrome conversion, can occur,[182] whereas others cannot. Thus, germination can be arrested at a stage requiring higher ψ_{soil} to allow further water uptake, metabolism, and cell expansion to occur.[183] Once embryonic expansion begins and the seed coat is broken, however, further growth can occur at lower ψ_{soil} than that which prevented germination. This is due at least partly to the large decrease in i_{seed} and the decrease in the turgor levels required for growth — both due to breakage of the seed coat. Prolonged exposure to a ψ_{soil} that will not allow germination can lead to secondary dormancy.[179]

Germination rates may be lowered by low ψ_{seed} through reduction or alteration of physiological processes which initiate germination. The best proof of this is the slowing of "escape time" (see Section II.C.1) from P_{fr} caused by lowered ψ_w.[113] The negative correlation between phytochrome escape time and germination rate suggests that there is a positive correlation between the rate of germination-triggering processes and the probability of germination, whether altered by temperature, ψ_{soil}, or other factors.

In *Avena fatua* lowered ψ_{seed} causes pronounced inhibition of germination by light.[184] Suppression of germination by light in these seeds was accompanied by a reduction in the rate of increase of ψ_{seed}, as compared with similar seeds germinated in darkness. Inhibition of germination by light at low ψ_{seed} may be due to the increased efficiency of P_{fr} conversion to P_r relative to that of P_r to P_{fr}. No interaction has been found between water stress and response of seeds to ethylene and allelochemicals.[185,186]

In *R. crispus* seeds water stress has no effect on the time of onset of secondary dormancy in seed populations; however, the rate of increase of secondary dormancy in the population was increased by decreased ψ_{soil} in some seed lots and unaffected in others.[113] When *R. crispus* seeds were preincubated in osmotica of various ψ_o in the light for 4 days and then transferred to water, the degree of secondary dormancy increased with decreasing ψ_o from 0 to 45% PEG (polyethylene glycol) 6000 (0 to -21 bars) and then decreased with decreasing ψ_o from 45 to 70% PEG 6000 solution.[179] Lowered ψ_w during the onset of secondary dormancy also reduced the effectiveness of various hormones in breaking secondary dormancy.

In a field environment, seeds are exposed to constantly changing ψ_{soil}, which results in a constantly cycling seed hydration that is broken only by germination. The effects of this cycling on the seed have been little studied, probably due to the complexities of the phenomenon. Rehydration of seeds previously imbibed and then dried often results in more rapid and higher levels of germination than if the seeds had not been allowed to dry. The apparent vigor has been explained by retention of physiological preparations for germination during dehydration, thus, "priming' the seed for germination. This phenomenon has been used extensively in horticultural crops and was reviewed by Hegarty.[187] One of the most impressive examples of the retention of a "priming" factor is that retention of phytochrome in the P_{fr} form for long periods of dehydration.[188] It has been shown, however, that germination takes approximately the same time to occur in once-imbibed seeds as in "primed" seed if the time of the first imbibition is considered.[189]

The effects of dehydration-rehydration upon seeds are complex and difficult to predict. In *R. crispus*, dehydration-rehydration at alternating temperatures can break dormancy and increase the speed of germination of seeds imbibed in darkness. However, at constant temperatures the treatment is ineffective (Table 1).[190] Alternating temperature before the dehydration/rehydration treatment was somewhat effective in causing germination of seeds subsequently imbibed at constant temperature, providing the first incubation lasted two or more days. If seeds were imbibed at constant temperature for two or more days before the dehydration/rehydration treatment, alternating temperatures in a subsequent imbibition were relatively ineffective in breaking dormancy. These data partially explain why seedlings of *R. crispus* appear rapidly in the summer after a drought-ending rain from seeds near the soil surface, whereas seeds of the same species located deep in the soil acquire a secondary dormancy.

Table 1
GERMINATION (%) OF *RUMEX CRISPUS* SEEDS IMBIBED IN DARKNESS AT ALTERNATING TEMPERATURES (18 H AT 25°C/10 H AT 10°C) OR AT A CONSTANT 25°C FOR VARIABLE PERIODS AND THEN DRIED FOR 24 H IN LIGHT UNDER THE SAME TEMPERATURE REGIMES. SEEDS WERE THEN GERMINATED FOR 7 DAYS IN DARKNESS UNDER THE ALTERNATING OR THE CONSTANT TEMPERATURE REGIME

	Days of imbibition before dehydration				
Treatments	0	1	2	2.5	3
Alternating temp during first imbibition					
Alternating temp during germination	54	96	96	96	94
Constant 25°C during germination	<1	1	15	53	53
Constant 25°C during first imbibition					
Alternating temp during germination	54	36	2	7	12
Constant 25°C during germination	<1	<1	0	0	0

Adapted from Vincent, E. M. and Cavers, P. B., *Can. J. Bot.*, 56, 2207, 1978. With permission.

5. Mineral Ions, pH, and Other Soil Factors

Very little is known of the physiological effects of the inorganic, nonosmotic, chemical aspects of soil chemistry, other than nitrate and nitrite (see below), on dormancy or germination. In general, metal cations have little or no effect on germination. For example, Ca^{2+} and Mn^{2+} have been reported to enhance, inhibit, or have no effect on seed germination, depending on the species.[191,192] In rice, Cu^{2+}, Hg^{2+}, and Ag^{2+} retarded breaking of dormancy, although neither these ions nor Na^+, K^+, Mg^{2+}, Fe^{2+}, Fe^{3+}, Co^{2+}, Ni^{2+}, or Zn^{2+} at concentrations up to 10 m*M* affected germination.[193] At high Ca^{2+}, Mg^{2+}, and Na^+ concentrations, inhibition of *Ericameria austrotexana* germination was due almost entirely to osmotic effects.[194] No extensive studies of metal cation effects on weed seed germination and dormancy have been conducted. The general ineffectiveness of these compounds, even at unnaturally high concentrations in the studies that have been conducted, suggests that the effort would not be rewarding.

Inorganic anions other than nitrate and nitrite have generally been reported to have no significant effects on germination. We are aware, however, of no literature on their effects on dormancy. Their lack of effect may be due to lack of penetration before germination.

Hydrogen ion has been found to have no effect on germination of seeds of several species within a range of pHs normally found in the soil.[195,196] Low pH has been reported, however, to enhance the effectiveness of nitrate and gibberellic acid in stimulating germination of *Lactuca sativa*[197] and of nitrite in breaking dormancy of *Oryza sativa*.[198]

In general, there is no evidence of substantial influences of inorganic ions, other than nitrate, nitrite, or hydrogen, on dormancy or germination. This is an area that deserves more attention, particularly with weed seeds, because they usually remain in contact with the soil much longer than the crop seeds with which most of the studies have been conducted.

6. Allelochemicals

Allelopathic interactions between weed and crop species are subjects of growing scientific interest (see Chapter 5). Although there is ample evidence of the involvement of allelochemicals in the growth and development of weed and crop species in agricultural situations, whether these compounds commonly affect germination of weeds is not clear.

Most allelochemicals studied have been phenolic compounds. Very little is known of the physiological mechanism of their inhibition of germination, although there is indirect evidence that some of these compounds have mechanisms of action differing from others.[199] For instance, inhibition of germination by *p*-coumaric acid and some of its immediate derivatives probably results from the inhibition of K^+ pumping and the resulting inability of embryonic cells to develop sufficient pressure potential to cause cellular expansion.[185] Coumarin, however, apparently affects germination through a different, but unknown, mechanism.

Most studies of the effects of allelopathic compounds on seed germination have been done in petri dishes with only one chemical at a time (e.g., Liebl and Worsham[200]). Although, in those studies, little evidence of effects on germination have been reported at chemical concentrations detected in the soil, these compounds might act synergistically by themselves or with environmental factors not found in a petri dish to inhibit germination. Synergistic interactions between various phenolic compounds in their inhibition of seed germination have been reported;[201-203] however, other studies have cast doubt on significance of these results.[185,199] No evidence of synergism between inhibition of germination by water stress and phenolic acids was found.[185]

III. LONGEVITY OF WEED SEEDS

The large reservoir of seeds in the soil, also termed the soil seed bank, may consist of thousands of weed seeds per square meter and is a source of weed seeds for future infestations (Figure 12).[204] Thompson and Grime[205] described two fundamental types of seed banks as being transient or persistent. The transient seed bank contains seeds that remain viable for only 1 year. The persistent seed bank contains seeds that do not germinate during the first year after production because many of them are either in primary or secondary dormancy. These seeds usually become more deeply buried, and are the major source of future weed infestations. Seeds of some weed species can remain viable in soils for longer than 20 years,[3,5,206] therefore, some seeds may remain dormant and viable for several years and then germinate to produce weed stands.

Several conditions affect seed longevity in soil. Since *in situ* germination is apparently the predominant fate of weed seeds, factors favoring germination should reduce seed longevity and persistance.[145] Weed seed persistance has been reduced by soil disturbance, such as tillage, which results in increased soil aeration or exposure of seeds to light and generally improved germination conditions.[207] Roberts and co-workers found that in the absence of reseeding, weed seeds in cultivated soil were reduced in number by about 25% per year.[208,209] In general, weed seeds buried near the soil surface lose viability more rapidly than seeds buried more deeply.[210] High soil temperatures also favor loss of seed viability.[211]

Studies of the longevity of weed seeds in undisturbed sites indicated that seeds of *Rumex crispus, Oenothera biennis,* and *Verbascum blattaria* were still viable after 80 years.[206] Seeds of *Chenopodium album* and *Ranunculus repens* remained viable for at least 20 years.[5] Dawson and Bruns[212] found that a few seeds of weedy grasses (*Echinochloa crus-galli, Setaria viridis,* and *S. lutescens*) remained viable in the soil for 13 years. Seeds with coats

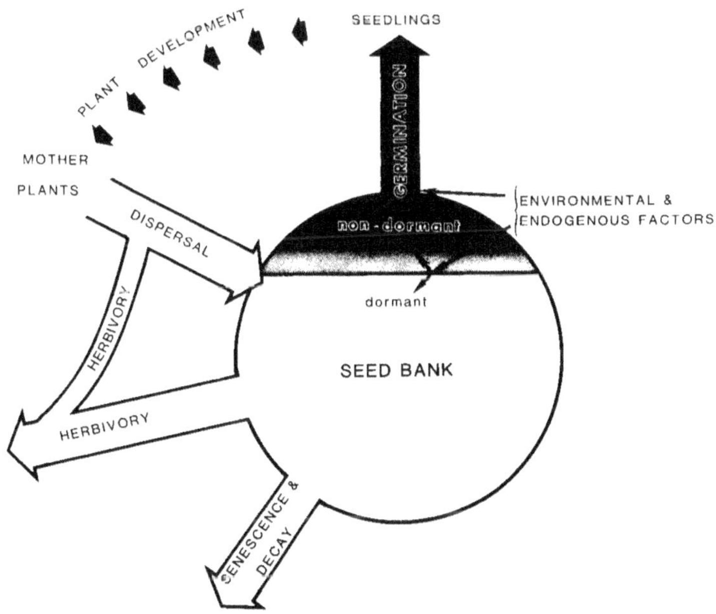

FIGURE 12. Diagrammatic flow chart for the dynamics of the population of seeds in the soil.

impermeable to water (*Abutilon theophrasti, Anoda cristata, Sesbania exaltata, Ipomoea turbinata*) comprised most of the longer-lived seeds at the $5^1/_2$ year point in a long-term seed longevity study in the warm, humid southern U.S.[4] Weed seeds also have the ability to survive for many years in the imbibed condition as well, however.

Villiers[213] hypothesized that hydrated, nongerminated seeds may extend their period of viability through self-repair or replacement of cellular components (DNA, enzymes, membranes), which may help to explain how weed seeds can remain viable in the soil for long periods. Villiers suggested that seeds normally incur damage (e.g., membrane disorganization), but the damage may be repaired or essential cell components replaced in the hydrated seeds during storage. Seeds dried below a critical level would not have sufficient moisture to support the metabolism necessary for the repairs. Membrane damage or failure apparently occurs to some extent during early germination because solutes leak from the embryo.[181,214] Ultrastructural evidence to support Villier's hypothesis of membrane damage and regeneration was obtained with *Lactuca sativa* and *Fraxinus* spp. seeds.[215]

Dormant *Avena fatua* seed tissues metabolize and have protein and lipid turnover rates similar to those of nondormant tissues.[83,216,217] Bewley and Black[10] reviewed reports that metabolic events also occurred in relatively dry (10 to 18% H_2O) seeds. Therefore, it is an attractive possibility that dormant weed seeds in the soil are able to repair damages or replace essential substances and remain viable. Further research is necessary to determine if these processes are responsible for weed seed longevity in nature.

IV. CHEMICAL MANIPULATION OF WEED SEED DORMANCY AND GERMINATION

Chemical manipulation of weed seed dormancy and germination in the field is one method by which weed seeds may be made more susceptible to weed control methods. It is important to consider chemicals that have broken dormancy or stimulated seed germination in a search for compounds that might be effective in the field (Table 2).

Table 2
LIST OF CHEMICALS WHICH BREAK DORMANCY OR STIMULATE GERMINATION OF SOME SEEDS

Growth regulators	Nitrogenous compounds
Gibberellins	Nitrate
Cytokinins	Nitrite
Ethylene	Hydroxylamine
Plant products	Thiourea
Fusicoccin	Sulfhydryl compounds
Cotylenol	Dithiothreitol
Cotylenin	2-Mercaptoethanol
Strigol	2,3-Dimercaptopropanol
Respiratory inhibitors	Various
Azide	Acetone
Cyanide	Ethanol
Malonate	Methanol
Hydrogen sulfide	Ethyl ether
Carbon monoxide	Chloroform
Sodium fluoride	Methylene blue
Iodoacetate	Carbon dioxide
Dinitrophenol	Phenols
L- and D-Chloramphenicol	Hydroxyquinoline
Hydroxylamine	Dimethylglyoxime
Oxidants	
Hypochlorite	
Oxygen	

From Bewley, J. D. and Black, M., *Physiology and Biochemistry of Seeds in Relation to Germination. Vol. 2. Viability, Dormancy and Environmental Control*, Springer-Verlag, New York, 1982, 375. With permission.

A. Hormonal Applications

Gibberellins are well-known germination stimulants of many weed seeds. In one study gibberellins induced germination of seeds in 5 of 19 weed species tested.[218] The responsive species were *Sinapis arvensis, Thlapsi arvense, Avena fatua, Loprula echinata,* and *Galeopsis tetrahit.* However, not all collections of the same species were responsive, which indicates a differential sensitivity to gibberellins. Paul et al.[219] reported that GA_3 stimulated the dark germination of fresh *Richardia scabra* seeds while kinetin applications did not. GA_3 also reversed the inhibitory effects of cold stratification upon the seeds. Thus, GA_3 was active on *R. scabra* seeds in either primary or secondary dormancy. Holm and Miller[53] determined seed germination responses of 11 weed species to several chemical treatments and found that 9 of the 11 were stimulated by GA_3, 7 were stimulated by thiourea, 4 were stimulated by kinetin, and 3 were induced to germinate by the ethylene-releasing compound ethephon (2-chloroethylphosphonic acid). The nine responsive species all were inhibited by ABA.

Cytokinins promote germination of *Xanthium* seeds (through expansion of cotyledons)[220] and kinetin and several other 6-substituted purines stimulate germination of *Striga* seeds.[221]

Ethylene also stimulates germination seeds of many species. Of 43 weed species that were tested for responses to ethylene, 9 were stimulated and 2 were inhibited by the hormone.[222] *S. asiatica,*[147] *X. pensylvanicum,*[154] and three other species[223] are also stimulated to germinate by ethylene.

Both gibberellins and cytokinins enhance germination of some weed seeds after seed coat scarification.[31,219,224] Thus, the penetration of the applied hormones to the site of action may be restricted by coats or seed coverings of some seeds. Preconditioning of seeds in hypochlorite, which may influence seed coat permeability, also enhances seed germination in response to applied hormones and regulators.[225,226] Thus, negative results with applied hormones or other chemicals may not be evidence for the lack of responsiveness of the seeds, because of the simple failure of the compound to reach the action sites.

ABA is a potent seed germination inhibitor.[227] Karssen[228,229] has shown that ABA does not prevent the initial events of *Chenopodium album* embryo expansion (early germination stage), but it does inhibit the embryo growth (later germination stage) necessary to penetrate the coverings over the embryo. Kahn and Samimy[224] reported similar results with *Rumex crispus* seeds. Karssen[230] also showed that ABA indirectly induces an ABA-independent secondary dormancy through germination inhibition of lettuce seeds. Thus, ABA may be primarily preventing germination through growth inhibition rather than maintaining dormancy. Some evidence suggests that ABA functions to arrest embryo growth during seed formation on the parent plant when seed tissues are still hydrated and assimilate storage is still occurring.[63] Thus, ABA may prevent precocious germination (vivipary) as discussed by Karssen.[231] ABA residues may remain in mature seeds after dispersal but may not be significant in maintaining dormancy. The ABA present could, however, prevent the embryo growth necessary for protrusion through the embryo coverings.

B. Nonhormonal Applications

One of the better-known nonhormonal chemical seed germination stimulants is nitrate, which is active on several species of weed seeds, particularly those requiring light or chilling.[198,232,233]

Nitrate is a commonly used agricultural chemical which stimulates seed germination of a wide range of weed species. Other reduced forms of nitrogen (e.g., ammonia, amino acids, or urea) have little or no effect on germination. Nitrite is more effective in causing germination than is nitrate, indicating that nitrate action may be due to reduction to nitrite.[233] This view is further supported by the finding that red light doubles the activity of nitrate reductase in lettuce seeds that are stimulated to germinate by light in the presence of nitrate.[234,235] In this system nitrate stimulates nitrate reductase activity in light or darkness. Unfortunately, no studies have been done to determine if nitrate reductase activities of seeds correlate with efficacy of nitrate in causing germination.

The mechanism by which nitrite (or nitrate) causes germination is unknown, but Roberts and Smith[84] have speculated that it involves the pentose phosphate respiratory pathway. They proposed that reduction of nitrite with NADPH makes NADP available for reduction by pentose pathway dehydrogenases and, thus, increases the activity of the pathway (see Section II.B.2). Several respiratory inhibitors (e.g., cyanide, azide) induce germination of some weed seeds. The mode of action of these compounds has been discussed in Section II.B.2.[51,84]

Fusicoccin, the plant growth promoter produced by the fungus *Fusicoccum amygdali*, also promotes germination of seeds of several species, including *Phacelia tanacetifolia*.[236,237] Its role in cell elongation has been associated with proton release from cells during cell wall loosening,[238] but this role has not been confirmed as an effector in seed germination. Indoleacetic acid (IAA), the cell elongation promotor, is generally considered ineffective as a germination inducer, although there is some evidence implicating it in the germination of nondormant seeds.[239] Thus, there is no evidence that a cell elongation factor triggers germination through cell elongation or expansion prior to germination.

Strigol, which was isolated from *Gossypium hirsutum* root exudates, is a very potent promoter of *Striga* spp. seed germination.[240] Strigol analogs are active as germination stimulants of germination of some other species of parasitic weed seeds.[241] However, stimulatory effects of strigol upon nonparasitic seeds have not yet been established. Other *Striga* spp.

seed germination stimulants (e.g., coumarin, 6-substituted purines) were discussed in a previous review article.[242]

Hypochlorite induces germination of parasite weed seeds (*Striga, Alectra*) and the weedy grass, *Avena fatua*.[225,243,244] It is likely that hypochlorite enhances germination through effects on seed coat permeability properties or by oxidation of inhibitors.

Recently, substituted phthalamides were reported to mimic GA_3 in stimulating seed germination of five weed species of nine that were tested.[245] The responsive species were *A. fatua, Brassica kaber, Thlapsi arvense,* and *Phacelia tanacetifolia.*

The fact that anesthetics (e.g., ethanol, ethyl ether, acetone) overcame seed dormancy in some weedy species, particularly grasses, is interesting in that these compounds are cell membrane effectors.[246] Taylorson and Hendricks[247] suggested that alteration of cell membrane properties by anesthetics may be responsible for their effects on seed dormancy and germination. This is particularly important since membranes have been implicated in phytochrome action[122] and have been suggested as sites of ethylene synthesis in the plant cell.[248]

A special problem for manipulation of germination involves hard seeds. In the laboratory, scarification with acids or physical means usually improves germination of the hard seeds. Growth regulators do not affect these seeds that do not imbibe water. It is interesting that French and Leather[249] were able to induce swelling of some weed seeds by exposing them to aldehydes of short-chain fatty acids. The treatments proved toxic as the seeds imbibed water, but did not germinate normally.

C. Chemical Methods of Reducing Dormant Weed Seed Populations in the Field

There is one example where a germination stimulant has been successfully used in the field to enhance weed control.[7] Ethylene applications induced *Striga asiatica* seed germination and reduced the number of viable seeds of this plant parasite in soils of North and South Carolina in the U.S. *S. asiatica* is an obligate root parasite of *Zea maize* and several other warm season grasses. The seeds normally do not germinate unless they are in close proximity of plant roots which exude a stimulant to *Striga* seed germination. Since ethylene stimulated germination of the seeds in soil in the absence of host roots, a very attractive control method was possible. Stimulation of *Striga* seed germination in the absence of the host is lethal to this parasite since attachment to the host root is essential to the survival of the obligate parasite. Those parasite seedlings that successfully achieve parasitism and survive are controlled by conventional herbicide methods. Methods for applying ethylene to the soil were developed by the USDA and up to 6000 hectares of *Striga*-infested fields were treated annually. Tests indicated that soil populations of *S. asiatica* seeds were reduced by up to 90% in the ethylene-treated fields.[250] Ethylene application to soil is an important tool in the program to combat this parasite in infested areas of the U.S.[7]

There are formidable problems to be solved in achieving significant reductions in the dormant weed seed bank in the soil, but results today indicate that these problems are not unsolvable. It is very unlikely that the entire weed seed bank or even seeds of one species in the bank in a given field or location will be induced to germinate by just a few treatments. Seed dormancy mechanisms are a well-established biological means of survival and it will be very difficult to overcome them. However, through a better understanding of the mechanisms controlling dormancy and germination, through the discovery of processes regulating dormancy and germination, and the development of the technology necessary to use dormancy and germination regulating treatments in agriculture, we may be able to make inroads in manipulating weed life cycles and, thereby, reduce their impact upon crops.

V. CONCLUDING REMARKS

Research with weed seeds has contributed significantly to current knowledge of seed dormancy and germination. *Avena fatua, Xanthium pensylvanicum, Amaranthus retroflexus, Chenopodium album,* and *Rumex crispus* are some weed species whose seeds have been the subjects of considerable physiological research. Although progress has been made, we still do not know how dormancy and germination are regulated in these seeds. However, the same is true for the nonweed, *Lactuca sativa,* perhaps the most extensively studied seed.

In most instances, research has described weed seed responses to environmental, chemical, or physical factors and has determined those that stimulate or inhibit germination. The biochemical or physiological bases for the responses remain poorly understood. The role of seed phytochrome in the control of germination is a notable exception in that the seed response to light is explained as due to the conversion of phytochrome to active or inactive forms by radiation. However, the mode of action of the active form of phytochrome in breaking dormancy remains unknown. The alteration of cell membrane permeability is an interesting theory of how dormancy-breaking is accomplished.[247] But even if this is the critical, primary event in dormancy release, the question still remains as to what biochemical, physiological, or cytological events are accomplished by the membrane permeability change.

It is evident that dormancy is not a simple process and is not due to a general "switching off" of metabolic activity. Both dormant and nondormant seeds may conduct many of the same metabolic processes. However, the increased participation of a specific pathway or shift to a particular balance in activities between different pathways may be important in the release from dormancy or stimulation of germination. For example, a particular balance between the cyanide-resistant, alternate pathway and the cyanide-susceptible cytochrome pathway of respiration was implicated in release of *X. pensylvanicum* dormancy.[92] Still, the means by which the shifts might cause the dormancy release is obscure. Also, there is no clear evidence that dormancy is due to a general deficiency of growth hormones or excess of growth inhibitors, although addition of hormones to seeds often alters dormancy or promotes or inhibits germination.[10]

Piercing of seed coats or coat removal induces germination of many dormant seeds. Coat removal is one of the most effective methods of stimulating germination. The means by which intact coats or other tissues of the seed envelope affect dormancy and restrict germination have been described, but the physiological mechanisms are incompletely understood. The integrity of the seed coverings are very important, particularly in the case of water-impermeable seed coats, to the persistence of many weed seeds in the dormant condition in the soil.

It is becoming clear that there are multiple bases for dormancy even within a single species. Variability in time of germination for individual seeds within a population is common among weed seeds and may be due to genetic or phenotypic diversity. Research with genetically pure lines of *Avena fatua* has pointed out the importance of genetic diversity in germination expression. Of four genetically pure dormant lines isolated from *A. fatua,* three were induced to germinate by azide and one was not.[251] This diversity within a species resulted in at least two different blocks to germination and illustrates that one stimulus may not be active on all dormant seeds even within a species.

The expression of those genes controlling germination can also be modified by the environment. For example, *Portulaca oleracea* seeds, produced under different photoperiods or different qualities of light, may exhibit different dormancy levels.[252]

Further studies with genetically pure lines of weed seeds or with populations of seeds that have like germination requirements will help to determine physiological bases for dormancy breaking and germination induction. It is important to recognize that germination variations exist widely among weed seeds, but it is also important to reduce this complex situation into component parts, when possible, to facilitate systematic studies of dormancy and germination.

Also, to facilitate systematic investigations, distinction should be made between breaking of dormancy and stimulation of germination. Often researchers consider "breaking of dormancy" and "stimulation of germination" as synonymous terms. This is understandable because breaking of dormancy usually results in germination. The inability of a seed to germinate is normally the only indicator of dormancy. However, there are situations where dormancy is broken and seeds changed from a nongerminable to a germinable condition without germination occurring.

Karssen described distinct germination stages in *Chenopodium album* with the first stage as embryo expansion and the last stage as radicle protrusion through coverings over the embryo.[229] Similar stages were noted in *Rumex crispus*[224] and *Amaranthus retroflexus*.[263] Since in *C. album* applied ABA did not prevent initiation of embryo expansion but did inhibit radicle protrusion, Karssen concluded that ABA did not inhibit breaking of dormancy but did inhibit germination.[229] Thus, in some seeds, it should be possible to differentiate among germination stages and determine effects of different variables upon each. It may then be possible to determine factors that affect early or late germination stages and identify those factors that control dormancy release or germination. These investigations would be important to a better understanding of the physiology of weed seeds in nature, since seeds may break dormancy in the soil but not germinate because of some unfavorable condition and then pass into secondary dormancy.

Cytochemical techniques should be utilized to determine the sequence of biochemical and cytological events that occur in the embryo and surrounding tissues from breaking of seed dormancy to germination. Attention should be centered upon those events occurring in the embryonic growth axis and in the coverings directly over the axis. Since the release from dormancy ultimately depends upon the ability of the embryo to initiate expansion, and since germination depends upon embryo expansion and growth sufficient for protrusion through the coverings, it is reasonable that discovery of processes involved in the dormancy and germination processes will most likely result from cytochemical studies of these areas at the ultrastructural level.

ACKNOWLEDGMENTS

The authors express gratitude to Drs. J. D. Bewley, R. E. Hoagland, G. R. Leather, H. R. Leffler, H. S. Mayeux, Jr., R. B. Taylorson, and K. C. Vaughn for their helpful reviews of an earlier manuscript draft. The authors, however, take full responsibility for statements made in the published manuscript.

REFERENCES

1. **Chancellor, R. J.,** Dormancy in weed seeds, *Outlook Agric.*, 11, 87, 1982.
2. **Roberts, E. H.,** The interaction of environmental factors controlling loss of dormancy in seeds, *Ann. Appl. Biol.*, 98, 552, 1981.
3. **Chepil, W. S.,** Germination of weed seeds. I. Longevity, periodicity of germination, and vitality of seeds in cultivated soil, *Sci. Agric.*, 26, 307, 1946.
4. **Egley, G. H. and Chandler, J. M.,** Longevity of weed seeds after 5.5 years in the Stoneville 50-year buried-seed study, *Weed Sci.*, 31, 264, 1983.
5. **Lewis, J.,** Longevity of crop and weed seeds: survival after 20 years in soil, *Weed Res.*, 13, 179, 1973.
6. **Harper, J. L.,** The ecological significance of dormancy and its significance in weed control, *Proc. Int. Conf. Crop Prot.*, 4, 415, 1959.
7. **Egley, G. H.,** Ethylene stimulation of weed seed germination, *Agric. For. Bull. Univ. Alberta*, 5, 13, 1982.
8. **Chancellor, R. J.,** The manipulation of weed behaviour for control purposes, *Philos. Trans. R. Soc. London Ser.* B, 295, 103, 1981.

9. **Amen, R. D.,** A model of seed dormancy, *Bot. Rev.*, 34, 1, 1968.
10. **Bewley, J. D. and Black, M.,** *Physiology and Biochemistry of Seeds in Relation to Germination. Vol. 2. Viability, Dormancy and Environmental Control*, Springer-Verlag, New York, 1982, 375.
11. **Karssen, C. M.,** Seasonal patterns of dormancy in weed seed, in *The Physiology and Biochemistry of Seed Development, Dormancy and Germination*, Khan, A. A., Ed., Elsevier, New York, 1982, 243.
12. **Sussex, I. M.,** Dormancy and development, in *Dormancy and Developmental Arrest: Experimental Analysis in Plants and Animals*, Clutter, M. E., Ed., Academic Press, New York, 1978, 167.
13. **Werker, E.,** Seed dormancy as explained by the anatomy of embryo envelopes, *Isr. J. Bot.*, 29, 22, 1980/81.
14. **Rolston, M. P.,** Water impermeable seed dormancy, *Bot. Rev.*, 44, 365, 1978.
15. **Ballard, L. A. T.,** Physical barriers to germination, *Seed Sci. Technol.*, 1, 285, 1973.
16. **Egley, G. H., Paul, R. N., Jr., Vaughn, K. C., and Duke, S. O.,** Role of peroxidase in the development of water-impermeable seed coats in *Sida spinosa* L., *Planta*, 157, 224, 1983.
17. **Marbach, I. and Mayer, A. M.,** Changes in catechol oxidase and permeability to water in seed coats of *Pisum elatius* during seed development and maturation, *Plant Physiol.*, 56, 93, 1975.
18. **Egley, G. H.,** Seed coat impermeability and germination of showy crotalaria (*Crotalaria spectabilis*) seeds, *Weed Sci.*, 27, 355, 1979.
19. **Hyde, E. O. C.,** The function of the hilum in some *Papilionaceae* in relation to ripening of the seed and the permeability of the testa, *Ann. Bot.*, 18, 241, 1954.
20. **Egley, G. H., Paul, R. N., Jr., Duke, S. O., and Vaughn, K. C.,** Peroxidase involvement in lignin formation in impermeable seed coats of weedy leguminous and malvaceous species, *Plant. Physiol. Annu. Suppl.*, 72, 87, 1983.
21. **Dell, B.,** Structure and function of the strophiolar plug in seeds of *Albizia lophantha*, *Am. J. Bot.*, 67, 556, 1980.
22. **LaCroix, L. J. and Staniforth, D. W.,** Seed dormancy in velvetleaf, *Weeds*, 12, 171, 1964.
23. **Egley, G. H. and Paul, R. N., Jr.,** Morphological observations on the early imbibition of water by *Sida spinosa* (Malvaceae) seed, *Am. J. Bot.*, 68, 1056, 1981.
24. **Egley, G. H. and Paul, R. N., Jr.,** Development, structure and function of subpalisade cells in water impermeable *Sida spinosa* seeds, *Am. J. Bot.*, 69, 1402, 1982.
25. **Egley, G. H. and Paul, R. N, Jr.,** unpublished data, 1983.
26. **Quinlivan, B. J.,** Seed coat impermeability in legumes, *J. Aust. Inst. Agric. Sci.*, 37, 183, 1971.
27. **Porter, N. G. and Wareing, P. F.,** The role of the oxygen permeability of the seed coat in the dormancy of seed of *Xanthium pensylvanicum* Wallr., *J. Exp. Bot.*, 25, 583, 1974.
28. **Witztum, A., Gutterman, Y., and Evenari, M.,** Integumentary mucilage as an oxygen barrier during germination of *Blepharis persica* (Burm.) Kuntze, *Bot. Gaz.*, 130, 238, 1969.
29. **Coumans, M., Côme, D., and Gaspar, T.,** Stabilized dormancy in sugar beet fruits. I. Seed coats as a physicochemical barrier to oxygen, *Bot. Gaz.*, 137, 274, 1976.
30. **Côme, D. and Tissaoui, T.,** Interrelated effects of imbibition, temperature and oxygen on seed germination, in *Seed Ecology*, Heydecker, W., Ed., Butterworths, London, 1973, 157.
31. **Egley, G. H.,** Influence of the seed envelope and growth regulators upon seed dormancy in witchweed (*Striga lutea* Lour.), *Ann. Bot.*, 36, 755, 1972.
32. **Ikuma, H. and Thimann, K. V.,** The role of the seed coats in germination of photosensitive lettuce seeds, *Plant Cell Physiol.*, 4, 169, 1963.
33. **Chen, S. S. C. and Thimann, K. V.,** Nature of seed dormancy in *Phacelia tanacetifolia*, *Science*, 153, 1537, 1966.
34. **Barton, L. V.,** Dormancy in seeds imposed by the seed coat, *Handbo. Plant Physiol.*, 15, 727, 1965.
35. **Pavlista, A. D. and Haber, A. H.,** Embryo expansion without protrusion in lettuce seeds, *Plant Physiol.*, 46, 636, 1970.
36. **Halmer, P., Bewley, J. D., and Thorpe, T. A.,** Degradation of the endosperm cell walls of *Lactuca sativa* L. cv. Grand Rapids. Timing of mobilization of soluble sugars, lipid and phytate, *Planta*, 139, 1, 1978.
37. **Psaras, G., Georghiou, K., and Mitrakos, K.,** Red-light-induced endosperm preparation for radicle protrusion of lettuce embryos, *Bot. Gaz.*, 142, 13, 1981.
38. **Bewley, J. D. and Halmer, P.,** Embryo-endosperm interactions in the hydrolysis of lettuce seed reserves, *Isr. J. Bot.*, 29, 118, 1980/81.
39. **Dunlap, J. R. and Morgan, P. W.,** Characterization of ethylene/gibberellic acid control of germination in *Lactuca sativa* L., *Plant Cell Physiol.*, 18, 561, 1977.
40. **Adkins, S. W. and Ross, J. D.,** Studies in wild oat dormancy. I. The role of ethylene in dormancy breakage and germination of wild oat seeds (*Avena fatua* L.), *Plant Physiol.*, 67, 358, 1981.
41. **Ho, T. D., Abrams, J., and Varner, J. E.,** Effect of ethylene on the release of α-amylase through cell walls of barley aleuron layers, *Plant Physiol.*, 69, 1128, 1982.

42. **Goeschl, J. D., Rappaport, L., and Pratt, H. K.,** Ethylene as a factor regulating the growth of pea epicotyls subjected to physical stress, *Plant Physiol.*, 41, 877, 1966.
43. **Esashi, Y. and Leopold, A. C.,** Physical force in dormancy and germination of *Xanthium* seeds, *Plant Physiol.*, 43, 871, 1968.
44. **Wareing, P. F.,** Hormonal regulation of seed dormancy — past, present and future, in *The Physiology and Biochemistry of Seed Development, Dormancy and Germination*, Khan, A. A., Ed., Elsevier, New York, 1982, 185.
45. **Ketring, D. L.,** Ethylene and seed germination, in *The Physiology and Biochemistry of Seed Dormancy and Germination*, Khan, A. A., Ed., Elsevier/North-Holland, New York, 1977, 157.
46. **Taylorson, R. B.,** Response of weed seeds to ethylene and related hydrocarbons, *Weed Sci.*, 27, 7, 1979.
47. **Spencer, M. and Olson, A. O.,** Ethylene production and lipid mobilization during germination of castor beans, *Nature (London)*, 205, 699, 1965.
48. **Ketring, D. L. and Morgan, P. W.,** Physiology of oil seeds. IV. Role of endogenous ethylene and inhibitory regulators during natural and induced afterripening of dormant Virginia-type peanut seeds, *Plant Physiol.*, 50, 382, 1972.
49. **Esashi, Y., Ohhara, Y., Kotaki, K., and Watanabe, K.,** Two C_2H_4-producing systems in cocklebur seeds, *Planta*, 129, 23, 1976.
50. **Rudnicki, R. M., Braun, J. W., and Khan, A. A.,** Low pressure and ethylene in lettuce seed germination, *Physiol. Plant*, 43, 189, 1978.
51. **Taylorson, R. B. and Hendricks, S. B.,** Dormancy in seeds, *Annu. Rev. Plant. Physiol.*, 28, 331, 1977.
52. **Jones, R. L. and Stoddart, J. L.,** Gibberellins and seed germination, in *The Physiology and Biochemistry of Seed Dormancy and Germination*, Khan, A. A., Ed., Elsevier/North-Holland, New York, 1977, 77.
53. **Holm, R. E. and Miller, M. R.,** Weed seed germination responses to chemical and physical treatments, *Weed Sci.*, 20, 150, 1972.
54. **Black, M.,** The role of endogenous hormones in germination and dormancy, *Isr. J. Bot.*, 29, 181, 1980/81.
55. **Black, M. and Naylor, J. M.,** Prevention of the onset of seed dormancy by gibberellic acid, *Nature (London)*, 184, 468, 1959.
56. **Williams, P. M., Bradbeer, J. W., Gaskin, P., and MacMillan, J.,** Studies in seed dormancy. VIII. The identification and determination of gibberellins A_1 and A_9 in seeds of *Corylus avellana* L., *Planta*, 117, 101, 1974.
57. **Paleg, L. G.,** Physiological effects of gibberellins, *Annu. Rev. Plant Physiol.*, 16, 291, 1965.
58. **Ilan, I. and Gepstein, S.,** Hormonal regulation of food reserve breakdown in germinating dicotyledonous seeds, *Isr. J. Bot.*, 29, 193, 1980/81.
59. **Brown, N. A. and Van Staden, J.,** The effect of stratification on the endogenous cytokinin levels of seed of *Protea compacta* and *Lencadendron daphnoides*, *Physiol. Plant.*, 28, 388, 1973.
60. **Van Staden, J. and Wareing, P. F.,** The effect of light on endogenous cytokinin levels in seeds of *Rumex obtusifolius*, *Planta*, 104, 126, 1972.
61. **Van Staden, J., Olatoye, S. T., and Hall, M. A.,** Effect of light and ethylene upon cytokinin levels in seed of *Spergula arvensis*, *J. Exp. Bot.*, 24, 662, 1973.
62. **Van Staden, J.,** Seeds and cytokinins, *Physiol. Plant.*, 58, 340, 1983.
63. **Walton, D. C.,** Does ABA play a role in seed germination?, *Isr. J. Bot.*, 29, 168, 1980/81.
64. **Trewavas, A. J.,** Growth substance sensitivity: the limiting factor in plant development, *Physiol. Plant.*, 55, 60, 1982.
65. **Schonbeck, M. W. and Egley, G. H.,** Changes in sensitivity of *Amaranthus retroflexus* L. seeds to ethylene during preincubation. I. Constant temperatures, *Plant Cell Environ.*, 4, 229, 1981.
66. **Schonbeck, M. W. and Egley, G. H.,** Changes in sensitivity of *Amaranthus retroflexus* L. seeds to ethylene during preincubation. II. Effects of alternating temperature and burial in soil, *Plant Cell Environ.*, 4, 237, 1981.
67. **Katoh, H. and Esashi, Y.,** Dormancy and impotency of cocklebur seeds. II. Phase sequence in germination process, *Plant Cell Physiol.*, 16, 697, 1975.
68. **Berrie, A. M. M., Buller, D., Don, R., and Parker, W.,** Possible role of volatile fatty acids and abscisic acid in the dormancy of oats, *Plant Physiol.*, 63, 758, 1979.
69. **Metzger, J. D. and Sebesta, D. K.,** Role of endogenous growth regulators in seed dormancy of *Avena fatua*. I. Shortchain fatty acids, *Plant Physiol.*, 70, 1480, 1982.
70. **Taylorson, R. B. and Hendricks, S. B.,** Anesthetic release of seed dormancy — an overview, *Isr. J. Bot.*, 29, 273, 1980/81.
71. **Wesson, G. and Wareing, P. F.,** The induction of light sensitivity in weed seeds by burial, *J. Exp. Bot.*, 20, 414, 1969.
72. **Holm, R. E.,** Volatile metabolites controlling germination in buried weed seeds, *Plant Physiol.*, 50, 293, 1972.

73. **Taylorson, R. B.**, Release of volatiles during accelerated after-ripening of seeds, *Seed Sci. Technol.*, 7, 369, 1979.
74. **Wareing, P. F. and Foda, H. A.**, Growth inhibitors and dormancy in *Xanthium* seed, *Physiol. Plant.*, 10, 266, 1957.
75. **Esashi, Y., Kotaki, K., and Ohhara, Y.**, Induction of cocklebur seed germination by anaerobiosis: a question about the "inhibitor hypothesis" of seed dormancy, *Planta*, 129, 109, 1976.
76. **Kust, C. A.**, A germination inhibitor in *Striga* seeds, *Weeds*, 14, 327, 1966.
77. **Chen, S. C. C.**, Application of a volatile radioactive tracer to detect metabolic activities in dry wild oat seeds, in *Dry Biological Systems*, Crowe, J. H. and Clegg, J. S., Eds., Academic Press, New York, 1978, 175.
78. **Mayer, A. M. and Marbach, I.**, Biochemistry of the transition from resting to germinating state in seeds, *Prog. Phytochem.*, 7, 95, 1981.
79. **Hecker, V. M., Bernhardt, D., and Dunger, H.**, Aktivierung der radikularen DNA-Synthese wahrend der Nachreife von *Agrostemma githago* — Samen, *Biol. Rundsch.*, 15, 308, 1977.
80. **Hecker, M. and Bernhardt, D.**, Proteinbiosynthesen in Dormanten und Nachgereifen Embryonen und Samen von *Agrostemma githago*, *Phytochemistry*, 15, 1105, 1976.
81. **Hecker, M., Stelter, C., and Kohler, K.-H.**, Methylation of RNA in dormant and after-ripened embryos of *Agrostemma githago*, *Biochem. Physiol. Pflanzen.*, 170, 227, 1976.
82. **de Klerk, G. J. and Kinskens, H. F.**, Protein synthesis in dormant and afterripened *Agrostemma githago* L. embryos, *Biochem. Physiol. Pflanzen.*, 174, 537, 1979.
83. **Chen, S. C. and Varner, J. E.**, Respiration and protein synthesis in dormant and after-ripened seeds of *Avena fatua*, *Plant Physiol.*, 46, 108, 1970.
84. **Roberts, E. H. and Smith, R. D.**, Dormancy and the pentose phosphate pathway, in *The Physiology and Biochemistry of Seed Dormancy and Germination*, Khan, A. A., Ed., Elsevier/North-Holland, New York, 1977, 385.
85. **Simpson, G. M.**, Metabolic regulation of dormancy in seeds — a case history of the wild oat (*Avena fatua*), in *Experimental Analysis in Plants and Animals*, Clutter, M. E., Ed., Academic Press, New York, 1978, 167.
86. **Esashi, Y., Komatsu, H., Ushizawa, R., and Sakai, Y.**, Breaking of secondary dormancy in cocklebur seeds by cyanide and azide in combination with C_2H_2 and O_2 and their effects on cytochrome and alternative respiratory pathways, *Aust. J. Plant Physiol.*, 9, 97, 1982.
87. **Esashi, Y., Satoh, M., Saijoh, K., and Satoh, S.**, Thermoperiodism mechanism in the germination of cocklebur seeds, *Plant Cell Physiol.*, 24, 17, 1983.
88. **Upadhyaya, M. K., Simpson, G. M., and Naylor, J. M.**, Levels of glucose 6 phosphate and 6-phosphogluconate dehydrogenases in the embryos and endosperms of some dormant and nondormant lines of *Avena fatua* during germination, *Can. J. Bot.*, 59, 1640, 1981.
89. **Fuerst, E. P., Upadhyaya, M. K., Simpson, G. M., Naylor, J. M., and Adkins, S. W.**, A study of the relationship between seed dormancy and pentose phosphate pathway activity in *Avena fatua*, *Can. J. Bot.*, 61, 667, 1983.
90. **Satoh, S. and Esashi, Y.**, No contribution of the pentose phosphate pathway in dormancy-breaking of cocklebur seeds, *Physiol. Plant.*, 48, 243, 1980.
91. **Upadhyaya, M. K., Naylor, J. M., and Simpson, G. M.**, The physiological basis of seed dormancy in *Avena fatua*. II. On the involvement of alternative respiration in the stimulation of germination by sodium azide, *Physiol. Plant.*, 58, 119, 1983.
92. **Esashi, Y., Kusuyama, K., Tazaki, S., and Ishihara, N.**, Necessity of a balance between CN-sensitive and CN-resistant respiration for the germination of cocklebur seeds, *Plant Cell Physiol.*, 22, 65, 1981.
93. **Esashi, Y., Nobuyori, I., Saijoh, K., and Satoh, M.**, A bimodal germination response to temperature in cocklebur seeds. I. Cyanide-sensitive and cyanide-resistant respirations, *Plant Cell Environ.*, 6, 47, 1983.
94. **Esashi, Y., Nagayama, T., Satoh, M., and Saijoh, K.**, A bimodal germination response to temperature in cocklebur seeds. II. ATP and adenylate pool, *Plant Cell Environ.*, 6, 55, 1982.
95. **Black, M.**, Light-controlled germination of seeds, *Symp. Soc. Exp. Biol.*, 23, 193, 1969.
96. **Mayer, A. M. and Poljakoff-Mayber, A.**, *The Germination of Seeds*, 2nd ed., Pergamon Press, Oxford, 1975, 35.
97. **Woolley, J. T. and Stoller, E. W.**, Light penetration and light-induced seed germination in soil, *Plant Physiol.*, 61, 597, 1978.
98. **Frankland, B.**, Germination in shade, in *Plants and the Daylight Spectrum*, Smith, H., Ed., Academic Press, New York, 1981, 187.
99. **Karssen, C. M.**, Patterns of change in dormancy during burial of seeds in soil, *Isr. J. Bot.*, 29, 65, 1980/81.
100. **Smith, H.**, *Phytochrome and Photomorphogenesis*, McGraw-Hill, New York, 1975.

101. **Borthwick, H. A., Hendricks, S. B., Parker, M. W., Toole, E. H., and Toole, V. K.,** A reversible photoreaction controlling seed germination, *Proc. Natl. Acad. Sci. U.S.A.*, 38, 662, 1952.
102. **Frankland, B. and Letendre, R. J.,** Phytochrome and effects of shading on growth of woodland plants, *Photochem. Photobiol.*, 27, 223, 1978.
103. **Taylorson, R. B. and Borthwick, H. A.,** Light filtration by foliar canopies: significance for light-controlled weed seed germination, *Weed Sci.*, 17, 48, 1969.
104. **Stoutjesdijk, P.,** Spectral transmission curves of some types of leaf canopies with a note on seed germination, *Acta Bot. Neerl.*, 21, 165, 1972.
105. **Fenner, M.,** The inhibition of germination of *Bidens pilosa* seeds by leaf canopy shade in some natural vegetation types, *New Phytol.*, 84, 95, 1980.
106. **Gorski, T.,** Germination of seeds in the shadow of plants, *Physiol. Plant.*, 34, 342, 1975.
107. **Gorski, T., Gorska, K., and Rybicki, J.,** Studies on the germination of seeds under leaf canopy, *Flora*, 167, 289, 1978.
108. **Taylorson, R. B. and Hendricks, S. B.,** Changes in phytochrome expressed by germination of *Amaranthus retroflexus* L. seeds, *Plant Physiol.*, 47, 619, 1971.
109. **Taylorson, R. B.,** Changes in dormancy and viability of weed seeds in soils, *Weed Sci.*, 18, 265, 1970.
110. **Fenner, M.,** The induction of a light requirement in *Bidens pilosa* seeds by leaf canopy shade, *New Phytol.*, 84, 103, 1980.
111. **Duke, S. O., Egley, G. H., and Reger, B. J.,** Model for variable light sensitivity in imbibed dark-dormant seeds, *Plant Physiol.*, 59, 244, 1977.
112. **Karssen, C. M.,** The light promoted germination of the seeds of *Chenopodium album* L. I. The influence of the incubation time on quantity and rate of response to red light, *Acta Bot. Neerl.*, 16, 156, 1967.
113. **Duke, S. O.,** Interactions of seed water content with phytochrome-initiated germination of *Rumex crispus* (L.) seeds, *Plant Cell Physiol.*, 19, 1043, 1978.
114. **Taylorson, R. B. and Hendricks, S. B.,** Rehydration of phytochrome in imbibing seeds of *Amaranthus retroflexus* L., *Plant Physiol.*, 149, 663, 1972.
115. **Nikolaeva, M. G.,** *Physiology of Deep Dormancy in Seeds*, National Science Foundation, Washington, D.C., 1969.
116. **Taylorson, R. B. and Hendricks, S. B.,** Phytochrome transformation and action in seeds of *Rumex crispus* L. during secondary dormancy, *Plant Physiol.*, 52, 475, 1973.
117. **Karssen, C. M.,** The light promoted germination of the seeds of *Chenopodium album* L. V. Dark reactions regulating quantity and rate of the response to red light, *Acta Bot. Neerl.*, 19, 187, 1970.
118. **Duke, S. O.,** Significance of fluence-response data in phytochrome-initiated seed germination, *Photochem. Photobiol.*, 28, 383, 1978.
119. **Le Deunff, Y.,** Mise en evidence du phytochrome chez les semences de *Rumex crispus* L., *Physiol. Veg.*, 9, 201, 1971.
120. **Le Deunff, Y. and Chaussat, R.,** Étude de la dormance secondaire de semences chez *Rumex crispus* L., *Ann. Physiol. Veg. Univ. Bruxelles*, 10, 227, 1968.
121. **Totterdell, S. and Roberts, E. H.,** Effects of low temperature on the loss of innate dormancy and the development of induced dormancy in seeds of *Rumex obtusifolus* L. and *Rumex crispus* L., *Plant Cell Environ.*, 2, 131, 1979.
122. **Marme, D.,** Phytochrome: membranes as possible sites of primary action, *Annu. Rev. Plant Physiol.*, 28, 173, 1977.
123. **Bewley, J. D., Black, M., and Negbi, M.,** Immediate action of phytochrome in light-stimulated lettuce seeds, *Nature (London)*, 215, 648, 1967.
124. **Leung, D. W. M. and Bewley, J. D.,** Immediate phytochrome action in inducing α-galactosidase in lettuce seeds, *Nature (London)*, 289, 587, 1981.
125. **Carpita, N. C., Nabors, M. W., Ross, C. W., and Petretic, N. L.,** The growth physics and water relations of red-light-induced germination in lettuce seeds. III. Changes in the osmotic and pressure potential in the embryonic axes of red- and far-red-treated seeds, *Planta*, 144, 217, 1979.
126. **Carpita, N. C., Nabors, M. W., Ross, C. W., and Petretic, N. L.,** The growth physics and water relations of red-light-induced germination in lettuce seeds. IV. Biochemical changes in the embryonic axes of red- and far-red-treated seeds, *Planta*, 144, 225, 1979.
127. **Bartley, M. R. and Frankland, B.,** Analysis of the dual role of phytochrome in the photoinhibition of seed germination, *Nature (London)*, 300, 750, 1982.
128. **Hilton, J. R.,** An unusual effect of the far-red absorbing form of phytochrome: photoinhibition of seed germination in *Bromus sterilis* L., *Planta*, 155, 524, 1982.
129. **Ballard, L. A. T. and Lipp, A. E. G.,** Studies of dormancy in the seeds of subterranean clover (*Trifolium subterraneum*). III. Dormancy breaking by low concentrations of oxygen, *Aust. J. Biol. Sci.*, 22, 279, 1969.
130. **Tissaoui, T. and Côme, D.,** Breaking of dormancy of the apple embryo (*Pirus malus* L.) in the absence of oxygen and cold, *Planta*, 111, 315, 1973.

131. **Lonchamp, J. P. and Gora, M.,** Influence of oxygen deficiencies on the germination of weed seeds, *Oecol. Plant.*, 14, 121, 1979.
132. **Morinaga, T.,** Germination of seeds under water, *Am. J. Bot.*, 13, 126, 1926.
133. **Brennan, T., Willemsen, R., Rudd, T., and Frenkel, C.,** Interaction of oxygen and ethylene in the release of ragweed seeds from dormancy, *Bot. Gaz.*, 139, 46, 1978.
134. **Esashi, Y., Tsukada, Y., and Ohhara, Y.,** Interrelation between low temperature and anaerobiosis in the induction of germination of cocklebur seed, *Aust. J. Plant Physiol.*, 5, 337, 1978.
135. **Simmonds, J. A. and Simpson, G. M.,** Increased participation of pentose phosphate pathway in response to afterripening and gibberellic acid treatment in caryopses of *Avena fatua*, *Can. J. Bot.*, 49, 1833, 1971.
136. **Fujii, T. and Yokohama, Y.,** Physiology of light-requiring germination in *Eragrostis* seeds, *Plant Cell Physiol.*, 6, 135, 1965.
137. **Crawford, R. M. M.,** Tolerance of anoxia and ethanol metabolism in germinating seeds, *New Phytol.*, 79, 511, 1977.
138. **Bewley, J. D. and Black, M.,** *Physiology and Biochemistry of Seeds in Relation to Germination. Vol. 1, Development Germination and Growth,* Springer-Verlag, New York, 1978, 306.
139. **Brown, R.,** Germination, in *Plant Physiology, A Treatise, Vol. VIC,* Steward, F. C., Ed., Academic Press, New York, 1972, 3.
140. **Aldasoro, J. and Nicolas, G.,** Fermentative products and dark fixation during germination of seeds of *Cicer arietinum*, *Phytochemistry*, 19, 3, 1980.
141. **Kennedy, R. A., Barrett, S. C. H., VanderZee, D., and Rumpho, M. E.,** Germination and seedling growth under anaerobic conditions in *Echinochloa crus-galli* (barnyardgrass), *Plant Cell Environ.*, 3, 243, 1980.
142. **Rumpho, M. E. and Kennedy, R. A.,** Anaerobic metabolism in germinating seeds of *Echinochloa crus-galli* (barnyardgrass): metabolite and enzyme studies, *Plant Physiol.*, 68, 165, 1981.
143. **Kennedy, R. A., Rumpho, M. E., and VanderZee, D.,** Germination of *Echinochloa crus-galli* (barnyard grass) seeds under anaerobic conditions. Respiration and response to metabolic inhibitors, *Plant Physiol.*, 72, 787, 1983.
144. **Kidd, F.,** The controlling influence of carbon dioxide in the maturation, dormancy, and germination of seeds. I, *Proc. R. Soc. London*, 87, 408, 1914.
145. **Roberts, E. H.,** Dormancy: a factor affecting seed survival in the soil, in *Viability of Seeds,* Roberts, E. H., Ed., Syracuse University Press, Syracuse, N.Y., 1972, 321.
146. **Bibbey, R. O.,** Physiological studies of weed seed germination, *Plant Physiol.*, 23, 467, 1948.
147. **Egley, G. H. and Dale, J. E.,** Ethylene, 2-chloroethylphosphonic acid, and witchweed germination, *Weed Sci.*, 18, 586, 1970.
148. **Toole, V. K., Bailey, W. K., and Toole, E. H.,** Factors influencing dormancy of peanut seeds, *Plant Physiol.*, 39, 822, 1964.
149. **Ballard, L. A. T.,** Studies of dormancy in the seeds of subterranean clover (*Trifolium subterraneum* L.). I. Breaking of dormancy by carbon dioxide and by activated carbon, *Aust. J. Biol. Sci.*, 11, 246, 1958.
150. **Schonbeck, M. W. and Egley, G. H.,** Phase-sequence of redroot pigweed seed germination responses to ethylene and other stimuli, *Plant Physiol.*, 68, 175, 1981.
151. **Hart, J. W. and Berrie, A. M. M.,** The germination of *Avena fatua* under different gaseous environments, *Physiol. Plant.*, 19, 1020, 1966.
152. **Jones, J. F. and Hall, M. H.,** Studies on the requirement for carbon dioxide and ethylene for germination of *Spergula arvensis* seeds, *Plant Sci. Lett.*, 16, 87, 1979.
153. **Negm, F. B., Smith, O. E., and Kumamoto, J.,** Interaction of carbon dioxide and ethylene in overcoming thermodormancy of lettuce seeds, *Plant Physiol*,. 49, 869, 1972.
154. **Esashi, Y. and Katoh, H.,** Dormancy and impotency of cocklebur seeds. III. CO_2- and C_2H_4-dependent growth of the embryonic axis and cotyledon segments, *Plant Cell Physiol.*, 16, 707, 1975.
155. **Esashi, Y. and Leopold, A. C.,** Dormancy regulation in subterranean clover seeds by ethylene, *Plant Physiol.*, 44, 1470, 1969.
156. **Egley, G. H.,** The ethylene-stimulated germination of light-sensitive common purslane seeds, *Plant Physiol.*, 61 (Suppl.), 16, 1978.
157. **Esashi, Y., Okazaki, M., Yanai, N., and Hishinuma, K.,** Control of the germination of secondary dormant cocklebur seeds by various germination stimulants, *Plant Cell Physiol.*, 19, 1497, 1978.
158. **Burdett, A. N.,** Antagonistic effects of high and low temperature pretreatments on the germination and pregermination ethylene synthesis of lettuce seeds, *Plant Physiol.*, 50, 201, 1972.
159. **Kao, C. H. and Yang, S. F.,** Light inhibition of the conversion of 1-aminocyclopropane-1-carboxylic acid to ethylene in leaves is mediated through carbon dioxide, *Planta,* 155, 261, 1982.
160. **Satoh, S. and Esashi, Y.,** Effects of a-aminoisobutyric acid and D- and L-amino acids on ethylene production and content of 1-aminocyclopropane-1-carboxylic acid in cotyledonary segments of cocklebur seeds, *Physiol. Plant.*, 54, 147, 1982.

161. **Smith, K. A. and Restall, S. W. F.,** The occurrence of ethylene in anaerobic soil, *J. Soil Sci.*, 22, 430, 1971.
162. **Hageseth, G. T. and Joyner, R. D.,** Kinetics and thermodynamics of isothermal seed germination, *J. Theor. Biol.*, 53, 51, 1975.
163. **Labouriau, L. G.,** Seed germination as a thermobiological problem, *Radiat. Environ. Biophys.*, 15, 345, 1978.
164. **Nishiyama, I.,** A break in the Arrhenius plot of germination activity in rice seeds, *Plant Cell Physiol.*, 16, 533, 1975.
165. **Duke, S. H., Schrader, L. E., and Miller, M. G.,** Low temperature effects on soybean (*Glycine max* [L.] Merr. cv. Wells) mitochondrial respiration and several dehydrogenases during imbibition and germination, *Plant Physiol.*, 60, 716, 1977.
166. **Roberts, H. A. and Neilson, J. E.,** Seasonal changes in the temperature requirements for germination of buried seeds of *Aphanes arvensis* L., *New Phytol.*, 92, 159, 1982.
167. **Taylorson, R. B. and Hendricks, S. B.,** Interactions of light and a temperature shift on seed germination, *Plant Physiol.*, 49, 127, 1972.
168. **Thompson, K., Grime, J. P., and Mason, G.,** Seed germination in response to diurnal fluctuations of temperature, *Nature (London)*, 267, 147, 1977.
169. **Totterdell, S. and Roberts, E. H.,** Characteristics of alternating temperatures which stimulate loss of dormancy in seeds of *Rumex obtusifolius* L. and *Rumex crispus* L., *Plant Cell Environ.*, 3, 3, 1980.
170. **Taylorson, R. B. and Hendricks, S. B.,** Phytochrome control of germination of *Rumex crispus* L. seeds induced by temperature shifts, *Plant Physiol.*, 50, 645, 1972.
171. **Goloff, A. H., Jr. and Bazzaz, F. A.,** A germination model for natural seed populations, *J. Theor. Biol.*, 52, 129, 1975.
172. **Toole, V. K.,** Light and temperature control of germination in *Agropyron smithii* seeds, *Plant Cell Physiol.*, 17, 1263, 1972.
173. **Hand, D. J., Craig, G., Takaki, M., and Kendrick, R. E.,** Interaction of light and temperature on seed germination of *Rumex obtusifolius* L., *Planta*, 156, 457, 1982.
174. **Roberts, E. H. and Totterdell, S.,** Seed dormancy in *Rumex* species in response to environmental factors, *Plant Cell Environ.*, 4, 97, 1981.
175. **Shukla, S. P.,** Nature of dormancy and germination regulation in seeds of *Portulaca quadrifida* L., *Jpn. J. Ecol.*, 22, 12, 1972.
176. **Taylorson, R. B. and Brown, M. M.,** Accelerated after-ripening for overcoming seed dormancy in grass weeds, *Weed Sci.*, 25, 473, 1977.
177. **Taylorson, R. B. and Hendricks, S. B.,** Action of phytochrome during prechilling of *Amaranthus retroflexus* L. seeds, *Plant Physiol.*, 44, 821, 1969.
178. **Vincent, E. M. and Roberts, E. H.,** The interaction of light, nitrate and alternating temperature in promoting the germination of dormant seeds of common weed species, *Seed Sci. Technol.*, 56, 659, 1977.
179. **Samimy, C. and Khan, A. A.,** Secondary dormancy, growth regulator effects and embryo growth potential in curly dock (*Rumex crispus*) seeds, *Weed Sci.*, 31, 153, 1983.
180. **McDonough, W. T.,** Water potentials of seeds of *Bromus inermus* and *Medicago sativa* imbibed on media of various osmotic potentials, *Can. J. Bot.*, 54, 1997, 1976.
181. **Duke, S. H. and Kakefuda, G.,** Role of the testa in preventing cellular rupture during imbibition of legume seeds, *Plant Physiol.*, 67, 447, 1981.
182. **McArthur, J. A.,** Light effects upon dry lettuce seeds, *Planta*, 144, 1, 1978.
183. **Loercher, L.,** Persistence of red light induction in lettuce seeds of varying hydration, *Plant Physiol.*, 53, 503, 1974.
184. **Hsiao, A. I. and Simpson, G. M.,** Dormancy studies in seed of *Avena fatua*. VII. The effects of light and variation in water regime on germination, *Can. J. Bot.*, 49, 1347, 1971.
185. **Duke, S. O., Williams, R. D., and Markhark, A. H., III,**Interaction of moisture stress and three phenolic compounds on lettuce seed germination, *Ann. Bot.*, 52, 923, 1983.
186. **Schonbeck, M. W. and Egley, G. H.,** Effects of temperature, water potential, and light on germination responses of redroot pigweed seeds to ethylene, *Plant Physiol.*, 65, 1149, 1980.
187. **Hegarty, T. W.,** The physiology of seed hydration and dehydration, and the relation between water stress and the control of germination: a review, *Plant Cell Environ.*, 1, 101, 1978.
188. **Vidaver, W. and Hsiao, A. J.,** Persistence of phytochrome-mediated germination control in lettuce seeds for 1 year following a single monochromatic light phase, *Can. J. Bot.*, 50, 687, 1972.
189. **Akalehiywot, T. and Bewley, J. D.,** Dessication of oat grains during and following germination, and its effects upon protein synthesis, *Can. J. Bot.*, 58, 2349, 1980.
190. **Vincent, E. M. and Cavers, P. B.,** The effects of wetting and drying on the subsequent germination of *Rumex crispus*, *Can. J. Bot.*, 56, 2207, 1978.
191. **Roberts, E. H.,** The effects of inorganic ions on dormancy in rice seeds, *Physiol. Plant.*, 16, 732, 1963.

192. **Loo, T. L. and Tang, Y. W.,** Growth stimulation by manganese sulphate, indolacetic acid and colchicine in the seed germination and early growth of several cultivated plants, *Am. J. Bot.*, 32, 106, 1945.
193. **Schutte, K. H. and Martin, P.,** The influence of certain micro-nutrient deficiencies on the germination rates of radish and millet, *Plant Soil*, 10, 299, 1959.
194. **Mayeux, H. S., Jr.,** Germination of false broomweed (*Ericameria austrotexana*) seed, *Weed Sci.*, 30, 597, 1982.
195. **Reynolds, T.,** pH restraints on lettuce fruit germination, *Ann. Bot.*, 39, 797, 1975.
196. **Stout, M. and Tolman, B.,** Factors affecting the germination of sugar-beet and other seeds, with special reference to the toxic effects of ammonia, *J. Agric. Res.*, 63, 687, 1941.
197. **Burdett, A. N.,** Two effects of prolonged far-red lights on the response of lettuce seeds to exogenous gibberellin, *Plant Physiol.*, 49, 531, 1972.
198. **Cohn, M. A., Butera, D. L., and Hughes, J. A.,** Seed dormancy in red rice. III. Response to nitrite, nitrate and ammonium ions, *Plant Physiol.*, 73, 381, 1983.
199. **Williams, R. D. and Hoagland, R. E.,** The effects of naturally occurring phenolic compounds on seed germination, *Weed Sci.*, 30, 206, 1982.
200. **Liebl, R. A. and Worsham, D. A.,** Inhibition of pitted morningglory (*Ipomoea lacunosa* L.) and certain other weed species by phytotoxic components of wheat (*Triticum aestivum* L.) straw, *J. Chem. Ecol.*, 19, 1027, 1983.
201. **Einhillig, F. A. and Rasmussen, J. A.,** Synergistic inhibitory effects of vanillic and *p*-hydroxybenzoic acids on radish and grain sorghum, *J. Chem. Ecol.*, 4, 425, 1978.
202. **Rasmussen, J. A. and Einhillig, F. A.,** Synergistic inhibitory effects of *p*-coumaric and ferulic acids on germination and growth of grain sorghum, *J. Chem. Ecol.*, 3, 197, 1977.
203. **Rasmussen, J. A. and Einhillig, F. A.,** Inhibitory effects of combinations of three phenolic acids on grain sorghum germination, *Plant Sci. Lett.*, 14, 69, 1979.
204. **Harper, J. L.,** *Population Biology of Plants*, Academic Press, New York, 1977, 892.
205. **Thompson, K. and Grime, J. P.,** Seasonal variation in the seed banks of herbaceous species in ten contrasting habitats, *J. Ecol.*, 67, 893, 1979.
206. **Darlington, H. T. and Steinbauer, G. P.,** The eighty-year period for Dr. Beal's seed viability experiment, *Am. J. Bot.*, 48, 321, 1961.
207. **Roberts, H. A. and Feast, P. M.,** Emergence and longevity of seeds of annual weeds in cultivated and undisturbed soil, *J. Appl. Ecol.*, 10, 133, 1973.
208. **Roberts, H. A. and Dawkins, P. A.,** Effect of cultivation on the numbers of viable weed seeds in soil, *Weed Res.*, 7, 190, 1967.
209. **Roberts, H. A. and Neilson, J. E.,** Changes in the soil seed bank of four long term crop/herbicide experiments, *J. Appl. Ecol.*, 18, 661, 1981.
210. **Toole, E. H.,** Final results of the Duvel buried seed experiment, *J. Agric. Res.*, 72, 201, 1946.
211. **Schafer, D. E. and Chilcote, D. P.,** Factors influencing persistence and depletion in buried seed populations. II. The effects of soil temperature and moisture, *Crop Sci.*, 10, 342, 1970.
212. **Dawson, J. H. and Bruns, V. F.,** Longevity of barnyard grass, green foxtail and yellow foxtail seeds in soil, *Weed Sci.*, 23, 437, 1975.
213. **Villiers, T. A.,** Seed aging: chromosome stability and extended viability of seeds stored fully imbibed, *Plant Physiol.*, 53, 875, 1974.
214. **Simon, E. W.,** Membranes in dry and imbibing seeds, in *Dry Biological Systems*, Crowe, J. H. and Clegg, J. S., Eds., Academic Press, New York, 1978, 205.
215. **Villiers, T. A. and Edgecumbe, D. J.,** On the cause of seed deterioration in dry storage, *Seed Sci. Technol.*, 3, 761, 1975.
216. **Cumming, A. C. and Osborne, D. J.,** Membrane turnover in imbibed and dormant embryos of the wild oat (*Avena fatua L.*). I. Protein turnover and membrane replacement, *Planta*, 139, 209, 1978.
217. **Cumming, A. C. and Osborne, D. J.,** Membrane turnover in imbibed dormant embryos of wild oat (*Avena fatua* L.). II. Phospholipid turnover and membrane replacement, *Planta*, 139, 219, 1978.
218. **Corns, W. G.,** Effects of gibberellin treatments on germination of various species of weed seeds, *Can. J. Plant Sci.*, 40, 47, 1960.
219. **Paul, K. B., Crayton, J. L., and Biswas, P. K.,** Germination behavior of Florida pusley seeds. II. Effects of germination-stimulating chemicals, *Weed Sci.*, 24, 349, 1976.
220. **Esashi, Y. and Leopold, A. C.,** Cotyledon expansion as a bioassay for cytokinins, *Plant Physiol.*, 44, 618, 1969.
221. **Worsham, A. D., Moreland, D. E., and Klingman, G. C.,** Stimulation of *Striga asiatica* (witchweed) seed germination by 6-substituted purines, *Science*, 130, 1654, 1959.
222. **Taylorson, R. B.,** Response of weed seeds to ethylene and related hydrocarbons, *Weed Sci.*, 27, 7, 1979.
223. **Olatoye, S. T. and Hall, M. A.,** Interaction of ethylene and light on dormant weed seeds, in *Seed Ecology*, Heydecker, W., Ed., Butterworths, London, 1973, 233.

224. **Khan, A. A. and Samimy, C.,** Hormones in relation to primary and secondary seed dormancy, in *The Physiology and Biochemistry of Seed Development, Dormancy and Germination,* Khan, A. A., Ed., Elsevier, New York, 1982, 203.
225. **Hsiao, A. I.,** The effect of sodium hypochlorite and gibberellic acid on seed dormancy and germination of wild oats (*Avena fatua*), *Can. J. Bot.,* 57, 1729, 1979.
226. **Hsiao, A. I.,** The effect of sodium hypochlorite, gibberellic acid and light on seed dormancy and germination of stinkweed and wild mustard, *Can. J. Plant Sci.,* 60, 643, 1980.
227. **Ho, D. T.,** The mechanism of abscisic acid during seed germination, in *The Physiology and Biochemistry of Seed Development, Dormancy and Germination,* Khan, A. A., Ed., Elsevier, New York, 1982, 299.
228. **Karssen, C. M.,** Two sites of hormonal action during germination of *Chenopodium album* seeds, *Physiol. Plant.,* 36, 264, 1976.
229. **Karssen, C. M.,** Uptake and effect of abscisic acid during induction and progress of radicle growth in seeds of *Chenopodium album, Physiol. Plant.,* 36, 259, 1976.
230. **Karssen, C. M.,** Indirect effect of abscisic acid on the induction of secondary dormancy in lettuce seeds, *Physiol. Plant.,* 54, 258, 1982.
231. **Karssen, C. M.,** The role of endogenous hormones during seed development and the onset of primary dormancy, in *Plant Growth Substances,* Wareing, P. F., Ed., Academic Press, New York, 1982, 623.
232. **Steinbauer, G. P. and Grigsby, B.,** Interaction of temperature, light and moistening agent in the germination of weed seeds, *Weeds,* 5, 175, 1957.
233. **Vincent, E. M. and Roberts, E. H.,** The influence of chilling, light and nitrate on the germination of dormant seeds of common weed species, *Seed Sci. Technol.,* 7, 3, 1979.
234. **Knypl, J. S.,** No effect of fusicoccin on phytochrome controlled nitrate reductase activity, *Plant Sci. Lett.,* 18, 319, 1980.
235. **Knypl, J. S.,** Hormonal control of nitrate assimilation: do phytohormones and phytochrome control the activity of nitrate reductase?, in *Nitrogen Assimilation of Plants,* Hewitt, E. J. and Cutting, C. V., Eds., Academic Press, London, 541.
236. **Cocucci, S., Ranieri, A. M., Morgutti, S., and Ciroli, F.,** The role of darkness, GA, and Fusicoccin (FC) in breaking photodormancy in *Phacelia tanacetifolia* seeds, *Physiol. Plant.,* 52, 177, 1981.
237. **Lado, P., Rasi-Caldogno, F., and Colombo, R.,** Promoting effect of fusicoccin on seed germination, *Physiol. Plant.,* 31, 149, 1974.
238. **Marre, E.,** Fusicoccin: a tool in plant physiology, *Annu. Rev. Plant. Physiol.,* 30, 273, 1979.
239. **Tillberg, E.,** Indoleacetic acid levels in *Phaseolus, Zea* and *Pinus* during seed germination, *Plant Physiol.,* 60, 317, 1977.
240. **Cook, C. E., Whichard, L. P., Wall, M. E., Egley, G. H., Coggon, P., Luham, P. A., and McPhail, A. T.,** Germination stimulants. II. The structure of strigol — a potent seed germination stimulant for witchweed (*Striga lutea* Lour.), *J. Am. Chem. Soc.,* 94, 6198, 1972.
241. **Visser, J. H. and Johnson, A. W.,** The effect of certain strigol analogues on the seed germination of *Alectra, S. Afr. J. Bot.,* 1, 75, 1982.
242. **Moreland, D. E., Egley, G. H., Worsham, A. D., and Monaco, T. J.,** Regulation of plant growth by constituents from higher plants, in *Natural Pest Control Agents,* Adv. Chem. Ser. No. 53, American Chemical Society, Washington, D.C., 1966, 112.
243. **Okonkwo, S. N. C. and Nwoke, F. I. O.,** Bleach-induced germination and breakage of dormancy of seeds of *Alectra vogelii, Physiol. Plant.,* 35, 175, 1975.
244. **Hsiao, A. I., Worsham, A. D., and Moreland, D. E.,** Effects of sodium hypochlorite and certain plant growth regulators on germination of witchweed (*Striga asiatica*) seeds, *Weed Sci.,* 29, 98, 1981.
245. **Metzger, J. D.,** Promotion of germination of dormant weed seeds by substituted phthalimides and gibberellic acid, *Weed Sci.,* 31, 285, 1983.
246. **Taylorson, R. B. and Hendricks, S. B.,** Overcoming dormancy in seeds with ethanol and other anesthetics, *Planta,* 145, 507, 1979.
247. **Taylorson, R. B. and Hendricks, S. B.,** Anesthetic release of seed dormancy — an overview, *Isr. J. Bot.,* 29, 273, 1980/81.
248. **Lieberman, M.,** Biosynthesis and action of ethylene, *Annu. Rev. Plant Physiol.,* 30, 533, 1979.
249. **French, R. C. and Leather, G. R.,** Screening of nonanal and related volatile flavor compounds on the germination of 18 species of weed seed, *J. Agric. Food Chem.,* 27, 828, 1979.
250. **Eplee, R. E.,** Ethylene: a witchweed seed germination stimulant, *Weed Sci.,* 23, 433, 1975.
251. **Upadhyaya, M. K., Naylor, J. M., and Simpson, G. M.,** The physiological basis of seed dormancy in *Avena fatua* L. I. Action of the respiratory inhibitors sodium azide and salicylhydroxamic acid, *Physiol. Plant.,* 54, 419, 1982.
252. **Gutterman, Y.,** Differences in the progeny due to daylength and hormone treatment on the mother plant, in *Seed Ecology,* Heydecker, W., Ed., Butterworths, London, 1973, 59.
253. **Egley, G. H. and Schonbeck, M. W.,** Carbon dioxide and ethylene phase-sequencing effects upon redroot pigweed seed germination in the dark in flasks, *Plant. Physiol. Annu. Suppl.,* 65, 42, 1981.

Chapter 3

ASEXUAL REPRODUCTION OF WEEDS

Jaime Kigel and Dov Koller

TABLE OF CONTENTS

I. INTRODUCTION

A. Concept of a Weed

Weeds are plant species that adapted to man-made habitats,[1] and are usually innocuous wild plants in their geographical center of origin.[2] They are unintentionally spread and perpetuated by the activities of man and his domesticated animals, including those that have unintentionally become domesticated. Most weeds are herbaceus, but some shrubs and trees, such as *Prosopis* spp., *Acacia* spp., and *Rubus* spp., and cacti, such as *Opuntia* spp., are extremely noxious weeds in certain habitats.

Weed species are detrimental to man's well-being to varying degrees. Research has concentrated mainly on their control and eradication, and relatively little has been done with respect to their developmental physiology. Yet, despite the total warfare declared against them, weeds still thrive. While in some cases this has been traced to development of tolerance or resistance to the chemicals and other agrotechnical means employed, in others survival is undoubtedly due to developmental strategies that provide protection of the original germplasm and, thereby, contributes to indefinite regeneration and reproduction. It is in this context that developmental aspects of weed physiology that are involved in spread, reproduction, and dispersal should be studied in depth. In this chapter we shall deal with strategies of spread and reproduction that involve asexual and vegetative processes. Various physiological and ecological aspects of asexual reproduction in weeds have recently been reviewed,[3-5] and information on the geographic distribution and economic importance of the world's worst weeds has been summarized.[6]

B. Role of Asexual Reproduction

The scientific designation of all forms of asexual reprduction is *apomixis*.[7] This term includes all methods of propagation that permit the exact reproduction of the genotype of the parent plant in its offspring without gene segregation and recombination. All genetically identical individuals that arise by apomictic processes are members of the same clone.[8] Genetic and evolutionary consequences of apomixis have been extensively discussed.[9-11] Broadly speaking, apomictic processes fall into two categories:

1. *Agamospermy*, which is the scientific term describing reproduction by means of seeds containing embryos that have arisen without meiosis and fertilization of the female gamete (egg cell)
2. *Vegetative reproduction* by any of the following means: (1) propagules that develop in inflorescences, in place of flowers (vivipary); (2) apical and axillary buds on specially modified shoots, or parts of shoots, and (3) adventitious buds on roots, stems, or leaves

The adaptive value of the genetic stability that results from apomixis obviously lies in the perpetuation of a highly successful genotype in a habitat to which it is specifically adapted. Genetic stability is also achieved by the numerous plant species that are self-fertilizers. The difference lies in the fact that self-fertilization is conducive to a high degree of homozygosity, whereas apomixis enables the perpetuation of heterozygosity. Perpetuation of a genotype, no matter how well adapted it is to its habitat, nevertheless, carries with it the danger of extinction by changes that may occur in the habitat that may make it less hospitable (physically, chemically, or biologically through competition or predation). Most, if not all self-fertilizing species exhibit a small but persistent proportion of outbreeders, and this undoubtedly contributes to a constant input of closely related, but different genotypes into the natural population. Similarly, most apomictic species are facultative and conserve some degree of sexual reproduction.[12] Presumably, somatic mutations also contribute a small additional source of variability.

The potential for rapid adaptation is probably greater in species that combine normal sexual reproduction with apomixis (agamospermy or vegetative reproduction), than in those relying on one or the other exclusively. Well-adapted parents that combine sexual and asexual reproductive strategies utilize their capacity for asexual reproduction to establish extensive clones with high potential for repeated attempts at diversity, by producing new genotypes from the same gene pool by sexual processes.[13] Therefore, maintenance of the potential for sexual reproduction in conjunction with asexual reproduction provides highly adaptable species complexes, capable of rapid evolution under highly selective pressure. When apomixis predominates, on the other hand, it usually generates complexes of ecotypes, each adapted to a specific environmental niche (mosaic adaptation). It, therefore, enables the persistence of specialists with a wide range of biotypes, without genetic intermixing.[14,15]

Apomixis, as well as self-fertilization, allows a single individual plant to colonize a new ecological niche. However, in contrast to self-fertilization, apomixis enables the colonization of new habitats by single individuals that are self-incompatible, dioecious, male-sterile, sterile hybrids, or infertile polyploids, as well as in situations restricting flowering (e.g., short growing season) or pollination (absence of vectors, low population density).[11,16]

Male-sterile forms of *Pennisetum clandestinum*, introduced from subtropical Africa, thrive in California, spreading by stolons and rhizomes.[17] In the genus *Oxalis*, the pentaploid *O. pes-caprae* and *O. corymbosa* are both sterile and survive by their capacity to reproduce by means of bulbils.[16] Apomixis also enhances the conservation and perpetuation of unique, or rare, successful heterozygous hybrids that may appear in populations of outbreeding species. In absence of apomixis, the existence of such successful individual hybrids may be ephemeral, because the genetic recombination that is inherent in the sexual process may break down its genetic complex. Furthermore, apomixis may allow a better-adapted genotype to increase its frequency in the population much more rapidly and, thus, achieve supremacy through intraspecific competition within a short time, whereas recombination breaks down even successful combinations. On the other hand, in absence of sexual reproduction, competition between closely related clones may lead to the eventual domination of the habitat by very few clones, or even by a single one. This has been reported, for instance, in *Pteridium aquilinum*.[18] Evidently, the future of such monoclonal populations becomes highly dependent on stability of the habitat and is endangered by changes in it.

Asexual reproduction in weeds in closely associated with polyploidy.[19-21] Polyploids are frequently sterile and can only persist by apomixis. Furthermore, polyploid hybrids between parents differing in their ecological requirements result from their greater range of ecological tolerance than their diploid parents, and if they are sterile, they persist by apomixis.[22]

1. Role of Agamospermy

Agamospermy is widespread among angiosperms.[12] Weed species in which agamospermy is the common method of seed production include *Taraxacum officinale* aggr.,[23] European species of *Rubus*,[24,25] and *Chondrilla juncea*.[26] Agamospermous representatives have been reported in several weed species of the Gramineae, such as *Paspalum dilatatum*, *Panicum maximum*, and *Chloris gayana*.[12,27] Agamospermy may be obligatory or facultative. In the former, all seeds are produced asexually (as in *T. officinale* aggr.), whereas in the latter a varying proportion of the seeds arise by sexual processes (as in *Hypericum perforatum*[28] and *R. fruticosus*[29]). Furthermore, in some agamospermous species, development of apomictic seeds requires pollination (*pseudogamy*). Pseudogamy has been found in species of *Rubus, Hypericum, Potentilla, Ranunculus,* and *Allium*. Nonpseudogamus genera are *Taraxacum, Hieracium,* and *Calamagrostis*.[12,30] Pollination in pseudogamy may act by providing a stimulus for embryonic development of the asexually produced seed, or it may provide the nucleus that fuses with the polar nuclei and, thus, initiates endosperm development. Pollination is required for endosperm formation in the majority of apomictic Graminaeae in which the embryo sac is aposporous (egg develops from a somatic cell in the nucellus or inner integument without meiosis) as in *Panicum maximum*.[31] In facultative agamospermy, the balance between asexually and sexually produced seeds may be controlled environmentally and vary according to season,[32,33] or photoperiod,[34-36] or it may be under endogenous control.[37]

There are a number of different cytological pathways by which agamospermy is achieved, and these are reviewed in detail by Nygren[12] and Heslop-Harrison.[30] However, there is no evidence that indicates that the pathway by which an apomictic seed develops has any consequences on the survival and propagation of the clone. From the point of view of asexual reproduction, the outstanding characteristic of agamospermy is the combination of genetic stability with the biological functionality of the seed. Furthermore, by circumventing the sexual process involved in seed production, agamospermy also provides means and opportunities for reproduction and dispersal by seed in cases where sexual processes are inhibited by environmental, biotic, or endogenous factors (e.g., unsuitable environment for the development of viable pollen, anther dehiscence, absence of pollinators, or sterile hybrids). Despite the apparently advantageous combination between dispersal by seed and conservation of the genotype, agamospermy does not appear to be widespread in weeds.

2. Role of Vegetative Reproduction

Vegetative reproduction offers means of spread and propagation without recourse to any reproductive process that involves flowering (except in cases of vivipary). It, thus, compensates for the advantages of seed dispersal, seed tolerance of adverse environmental conditions, and seed dormancy, by a capacity to achieve reproduction and dispersal without involving the complicated processes that start with flower initiation and culminate with seed set and dissemination.

Vegetative reproduction is, therefore, obligatory in all cases where flowering or any of its subsequent reproductive processes are inhibited. This condition appears to prevail in a number of troublesome aquatic weeds: e.g., sterile triploids of *Acorus calamus* in Europe;[38,39] female plants of the dioecious *Elodea canadensis* in Europe;[10] monomorphic populations of the tristylic *Eichhornia crassipes* in California[16,40] and elsewhere.[41] Analogous situations where vegetative reproduction compensates for self-incompatibility have been reported for other aquatic weeds, such as *Butomus umbellatus* in Europe and *Pistia stratiotes* in the tropics.[42] Tristily (trimorphic self-incompatibility) is also known in land plants. A well-known example in the tristylic genus *Oxalis*, whose short-styled forms of *O. pes-caprae* and *O. latifolia* have used their capacity for vegetative reproduction to become weeds outside their centers of distribution.[43-47]

Vegetative reproduction enables the continued survival of sterile hybrids that have inherited from one of their parents the capacity for vegetative reproduction, e.g., the accidental hybrids between the cultivated, male-sterile *Sorghum bicolor* and its weedy relatives *S. halepense* and *S. almum*, both perennials.[48] The ability of the infertile hybrids to reproduce vegetatively and, thus, survive virtually indefinitely increases the probability of autpolyploidy that may result in fertility, e.g., *Spartina townsendii*.[49]

Vegetative reproduction exhibits distinctive features by which it differs from reproduction by seeds. Reproduction by vegetative organs presents far less risky opportunities for establishment than by seeds and, thus, compensates to a large extent for the greater abundance of seeds. Organs and structures of vegetative reproduction are able to depend on supply of nutrients and water from the parent plant for a much longer period. Transport of labeled assimilates from parent to daughter plants developing on stolons or rhizomes has been demonstrated in *Cynodon dactylon*,[50] *Agropyron repens*,[51] *Poa pratensis*,[52] and *Ranunculus repens*.[53] Establishment of daughter plants is, thus, much less dependent on the immediate availability of such essential environmental factors as water, light intensity, and minerals. Parental support provides the means for making considerable growth until the essential factor(s) becomes available.

In *A. repens*,[54] *Trifolium repens*,[55] and stoloniferous species of *Viola*,[56] seedlings have a lower probability to become established plants, compared to daughter plants connected through stolons to the ''parent''plants. In *Rubus fruticosus*, survival of daughter plants originated from adventitious buds on creeping roots was much higher than that of seedlings.[29] Similarly, the effects of competition are far less severe, or even significant, to daughter plants that are still attached to their parents. In dense stands of *Hieracium floribundum* a seed has very low probability to produce an established seedling, and they provide considerably fewer individuals to the population than those arising from vegetative reproduction.[57]

Parents may supply their vegetatively produced offspring with readily available assimilates for immediate use for growth of the daughter plants to which they are connected, or for storage in perennating structures (e.g., bulbs, corms, tubers, rhizomes, etc.) for subsequent utilization by activated buds. As with seeds, food storage is usually associated with bud and root dormancy of the perennating units. If disconnected from the parent plant, these units may function as vegetative propagules with a capacity of independent existence for prolonged periods in an inactive (dormant) state. In this form they can undergo spatial dislocation by exogenous agents, as well as withstand to some extent unfavorable conditions, much like seeds. However, the vegetative propagules of several weeds are much less resistant to desiccation than seeds. Tubers of *Cyperus rotundus*[58] and *C. esculentus*,[59,60] as well as rhizome segments of *Cynodon dactylon*[60] and *A. repens*,[61] and root segments of *Cirsium arvense*[62] dry rapidly if exposed to the air. Viability of such propagules is usually lost at water contents lower than 15 to 20%, while seeds retain viability at much lower water contents.

As means of propagation, vegetative propagules differ from seeds mostly in quantitative terms. They are usually larger and heavier than seeds and contain more stored food and water. The greater content of stored food allows to the new sprouts a faster shoot growth and more rapid emergence from a greater depth in the soil, and against a greater mechanical resistance, as compared to seedlings. They have the ability to emerge from deeper soil layers, where (1) they are in an environment that is hospitable for activation and the ensuing growth, and (2) they are more protected not only from extreme environmental conditions such as frost, heat, or drought, but also from predators and attempts at their eradication by man.

In weeds that spread vegetatively, the dispersal of the vegetative propagules is controlled primarily by the parent plant, from which they may be distanced by growth processes (e.g., bulbs and tubers that are formed at the tip of creeping stolons in *O. latifolia* and *C. esculentus*, respectively). The magnitude of this primary dispersal is, therefore, determined by the growth accomplished by the spreading organs, and differs between species and according to environmental conditions. In *Allium ampeloprasum*[44] and *O. corymbosa*,[63] for instance, bulbils are formed in close proximity to the main bulb. The dispersal of vegetative propagules is secondarily controlled by exogenous factors that disturb their soil habitat and, at the same time, detach them from their parents. In perennial weeds the most prevalent of these factors that contribute to their persistence and spread is cultivation.

Vegetative propagules lack the anatomical and morphological features by which weeds and seed-containing propagules are able to take advantage of regularly occurring and, therefore, dependable, environmental agents of dispersal such as wind, water, animals, etc. The soil medium through which most vegetative propagules are usually dispersed offers far greater frictional resistance to motion, and in combination with the far greater weight and size of the propagules it inhibits their long-distance dispersal. As a result, propagation by such means tends to create monoclonal colonies, whose individual members may compete with each other for the same environmental supplies. It is not surprising, therefore, that species that produce such propagules at relatively short distances from their parents also exhibit a correlative inhibition of their bud activation that is released after (or as a result of) disconnection of the propagule from the parent. Disconnection may be due to abscission processes, natural senescence and decay of the connecting organs, death of the parent plant, natural mechanical disturbance such as water currents, or cultivation (see Section II.D). On the other hand, development of the offspring at or very near the site occupied by the parent plant may increase its potential for establishment and reproduction by exploiting a "safe site", or by diminishing interspecific competition.

A disadvantage of vegetative reproduction, from which reproduction by seed is immune, is the capacity of transmitting and perpetuating viral infection in the offspring. This may lead to progressive loss of vigor and eventual degeneration of the clone, as has happened in a number of horticultural and ornamental clone lines that were propagated vegetatively. This may open up hitherto untested possibilities for control of vegetatively reproducing weeds, by deliberately infecting them with species-specific viruses.

Organs involved in vegetative propagation are usually produced much earlier in ontogeny than flowering, since the latter takes place at the end of the growing season, or of the growth cycle of the individual shoot or of the entire plant. Frequently, such organs are produced as early as the seedling stage. In *Euphorbia esula*[64] and *C. arvense*[62,65] the seedling root produces adventitious buds a few weeks after germination, while in *Linaria vulgaris*[66] and *Chondrilla juncea*[67] this early capacity is realized if the seedling tops are destroyed. In *Cynodon dactylon*[68,69] and *Agropyron repens*[70] rhizomes are formed before flowering. Furthermore, perennial weeds such as *Typha latifolia*[71] and *E. esula*[64] that do not flower during their first year after germination are able to produce rhizomes and creeping roots, respectively, very early in their development. Vegetatively reproducing species are, therefore, far better equipped to withstand repeated disturbances that may delay or totally prevent the initiation of flowering or the consummation of seed production and dispersal. In this context, vegetative production of propagules by *vivipary* is exceptional since it requires the fulfillment of the flowering process and inflorescence formation. In vivipary, developing flowers are modified to form plantlets or bulbils that are capable of dissemination and rooting, thus, establishing new plants. Vivipary is common in Festucoid grasses (*Deschampsia caespitosa, D. alpina, Festuca* spp., *Poa* spp.) and in Liliaceae (*Allium* spp., *Agave* spp.), but it occurs also in a number of dicots (*Polygonum viviparum, Cardamine bulbifera, Saxifraga foliolosa*).[9] Viviparous species may fail entirely to produce seed, while in others the tendency to vivipary is under environmental control. In *D. caespitosa*[72] and in several species of *Poa*,[73,74] short photoperiods induce vivipary. However, this means of vegetative reproduction is quite rare in weeds. *Allium vineale*, a species that produces bulbils in the inflorescence, is probably the best known weed that reproduces by vivipary.[75,76]

3. The Balance between Vegetative Reproduction and Reproduction by Seed

In plants that reproduce vegetatively as well as by seed, a balance exists in the partitioning of resources allocated to the asexual and sexual processes. In many cases, an inverse correlation between these opposite methods of reproduction has been found.[20,77] In several weeds with vegetative means of reproduction, the flowering stage is reached but no seeds are produced (e.g., *Armoracia rusticana*, *Polygonum amphibium*, *P. cuspidum*,[20] or biotypes of *Cynodon dactylon*[68]) or, if produced, they may be nonviable (e.g., *Cyperus rotundus*[78,79]). In other weeds that produce viable seeds, vegetative means of reproduction contribute more new plants than the seed bank in the soil. This is especially the case for weeds of pastures or other closed and relatively stable habitats in which interspecific competition is very intense: e.g., *Hypericum perforatum*,[80] *Rumex acetosella*.[81] *Ranunculus repens*,[54] *Hieracium floribundum*,[57] However, occasional new seedlings from sexually produced seeds maintain the genetic diversity of the population under conditions of interclonal competition.[13] Nevertheless, vegetative reproduction may be far less important than by seed, as in a number of perennial *Plantago* species.[82,83] Thus, the relative importance of seed and vegetative reproduction varies widely for different weed species growing in apparently similar habitats.

A better understanding of the survival value of these opposite methods of reproduction might be achieved by analysis of resource allocation between both methods under stress, or under conditions limiting growth. *Tussilago farfara*,[84] *Solidago canadensis*,[85] *Rubus hispidus* and *R. trivialis*,[86] and *Fragaria virginiana*,[87] growing in populations of increasing density, tend to allocate a larger proportion of their biomass to seed production than to vegetative reproduction. However, this response is not common to all weeds. In *C. esculentus* biomass allocation to tubers was unaffected by plant density,[88] while in *C. rotundus* it was increased at the expense of inflorescence production.[89]

The relative effects of density are probably limited by the plasticity of the developmental patterns of each species. In *V. blanda* and *V. rostrata*, mean individual biomass, as well as the proportion of the population that produces seeds and stolons, decreases with increasing population density. However, both species have a ''threshold'' biomass for seed and stolon production that is unaffected by density of the population. Therefore, at higher densities more plants fail to reach the threshold and do not produce seeds or stolons.[90] In *Eichhornia crassipes*, on the other hand, stolon production is reduced by higher plant densities in the floating mat (80% at the edge vs. 5% at the center produce stolons), but flowering is increased. This developmental shift from asexual to sexual reproduction was ascribed to decreasing levels of nutrients in the pond, especially nitrogenated compounds.[91] However, in *T. farfara* growing in soils of low fertility, the relative allocation of biomass to rhizome growth is increased at the expense of seed production.[84]

The balance between seed and vegetative reproduction is affected also by the light environment. *Lysimachia nummularia* and *Galeodolon luteum* growing in the shade reproduce by stolons, and they flower only in full light.[20] In *Sagittaria trifolia* and *S. aginashi*, noxious weeds in paddy fields, a 50% reduction in solar radiation increased tuber and decreased seed production in the former, and practically eliminated it in the latter.[92] In *C. esculentus* no inflorescences were produced at low light intensity, but the plants continued to produce tubers and were able to reproduce vegetatively.[93] Thus, the balance between reproduction by seeds and by vegetative means can be modulated by different environmental factors. The physiological mechanisms involved in these effects have not been investigated in sufficient detail.

II. STRATEGIES OF VEGETATIVE SPREAD AND REPRODUCTION

A. Introduction

A *growing* plant occupies a progessively larger space in the air and in the soil environment. If and when the growing plant develops secondary shoots that are able to produce their own root system, it is also *spreading vegetatively*, as well as growing. Thus, formation of secondary shoots (from axillary or adventitious buds) with a capacity for adventitious rooting is a prerequisite for spreading.

A plant that *spreads* vegetatively by these means may also have the capacity to *reproduce* itself vegetatively. This happens when the rooted shoot, or specialized bud from which such a shoot originates, becomes physically or physiologically disconnected from the structure on which it has been formed and achieves an independent existence. Spread can, therefore, be defined as progressive occupation of new territory by a single individual, whose parts are potentially capable of existing independently if separated from each other. From this definition it is clear that unless and until the capacity for becoming detached is realized, the entire population of interconnected rooted shoots (or the specialized buds from which they will eventually develop) is physiologically a single integrated individual whose parts remain more or less interdependent.

Transformation of spread into reproduction requires disconnection, and the processes involved in this may be endogenous and autonomous, may depend on environmental factors, or may result from agricultural manipulation. In the case of economically valuable species, manipulation is intentional, whereas in the case of economically detrimental weeds it is entirely unintentional and to a great extent unavoidable. In many cases, vegetative reproduction is associated with and depends on the production of a specialized type of bud, a so-called *regeneration bud*. These buds may differ in their origin. They may be modifications of apical or lateral buds, or they may develop adventitiously on a variety of plant organs (roots, stems, scales, or leaves) during the normal course of development, or as a result of fragmentation or injury.

Regeneration buds are usually produced in association with underground (or submerged) perennating structures possessing long-term storage capacity (e.g., rhizomes, tubers, corms, bulbs, roots, etc.). Such structures may remain dormant during seasons unfavorable for growth and are resistant to adverse environmental factors. Formation of the secondary shoots that contribute to spread may be localized to specific morphological regions of the original shoot or root system. This localization may greatly modify the spatial relationships between the original shoot-root system and the secondary rooted shoots, from a concentrated clump at one extreme, to a diffuse, wide-spaced system at the other. This is obviously reflected in the spatial distribution of the population of individual clonal plants that had reproduced by becoming separated from each other. The entire situation is changed when the habitat is mechanically disturbed, either by cultivation or by more natural agents, such as water flow in aquatic environments.

The potential for regeneration is determined by different factors in nonadventitious buds (e.g., on rhizomes) and in adventitious ones (e.g., on roots). In the former the upper limit depends on the number of nodes per unit length (or segment length), whereas in the latter it depends on the meristematic and morphogenetic potential of the tissues. In roots, adventitious buds are frequently closely associated with protoxylem poles (see Section II.B.2), whose number may, therefore, play a role in the regeneration potential. However, morphogenetic "inhibitory fields"[94] may affect the longitudinal separation between such buds.

B. Morphology and Function in Spread and Reproduction by Shoots and Roots

1. Nonadventitious Buds

Secondary shoots that root adventitiously are commonly diageotropic. A striking exception is found in several species belonging to the genus *Rubus*, where specialized orthotropic branches reverse their geotropic response and their tips grow down until they strike and enter the ground, forming a "rooting boss", from which a short, negatively geotropic shoot arises.[95]

The diageotropic shoots can be broadly subdivided into two categories, based on their functional role. As usual, there are a variety of intermediate categories that form a broad spectrum and confound attempts at clear-cut, well-distinguished characterization.

Shoots that clearly belong in one category are characterized by very long, slender internodes, that may have diminutive, vestigial leaves. These shoots root adventitiously at specific nodes (or at all of them) and form erect, leafy shoot systems at their points of rooting, from their axillary, or apical, buds. Such diageotropic shoots are commonly termed "stolons" or "runners" and their morphology and developmental pattern characterize them as instruments of territorial invasion and spread. Typical examples are *Fragaria* spp., *Ranunculus repens*, *Potentilla reptans*, *Paspalum dilatatum*, *Oxalis corniculata*, and *Trifolium repens*, whose stolons run on the soil surface and their apical, or axillary, buds take root and are transformed into erect, leafy shoots. New stolons are formed in the leaf axils of these erect shoots.

At the other extreme is the type of diageotropic shoot that is characterized by short, to extremely compressed, but thick internodes that are very rich in stored food (mostly carbohydrate). The nodes bear scale-leaves and specialized buds that are capable of remaining dormant for extensive periods. In certain species a few of these buds become activated early and contribute to form a sparsely branched, subterranean shoot system. This type of diageotropic shoot behaves as a perennating organ and has a clear-cut function in maintaining an extended capacity for vegetative regeneration. Such shoots are usually subterranean and are called "rhizomes". Typical examples are *Sorghum halepense*, *Agropyron repens*, *Imperata cylindrica*, *Phragmites communis*, *Poa pratensis*, *Pteridium aquilinum*, *Equisetum arvenase*, and *Tussilago farfara*. In *Poa pratensis*, soil penetration is assisted by spiral movements (circumnutation) of the actively growing stolon apices.[96]

Stolons and rhizomes may have either a determinate or an indeterminate developmental pattern, depending on whether or not the apical bud is transformed from a diageotropic to an erect growth habit, either with autonomous regularity or with a seasonally induced one (see Section III.C).

In-between these well-defined categories are found plants with subterranean plagiotropic shoots that alternate regions with thin, long internodes, carrying vestigial leaves and lacking dormant regeneration buds as well as stored food, with regions of short internodes, carrying dormant regeneration buds and rich in stored food. This storage is either in the short internodes, which are, consequently, very thick and form what is known as a "tuber", or in thick scale leaves that form what is known as a "bulb". Such "mixed" subterranean shoots evidently combine the advantages of territorial spread with those of sustained capacity for vegetative regeneration and perennation. For instance, *Allium ampeloprasum* forms a perennial bulb, from which numerous stolons are produced in the axils of its fleshy leaf bases. These are terminated by bulbils (or "bulblets") that may remain for a number of years, but the stolons do not survive the hot, dry summers.[44] *O. latifolia* forms bulbs consisting of fleshy scale-leaves that contain stored food. The apical bud of these bulbs forms foliage leaves and an inflorescence. Lateral buds that form in the axils of the scales grow into typical stolons, with thin, elongated internodes and vestigial scale-leaves. The apices of these stolons form "primary bulbils" which may, in turn, form branch stolons with "secondary bulbils" at their tips. The stolons are ephemeral and do not survive the first winter. Each bulbil eventually forms a contractile root, which changes its depth and orientation in the soil.[45] In *Stachys palustris* and *Mentha arvensis*, subterranean diageotropic shoots are formed from axillary buds at the base of the erect shoot. The proximal regions of these shoots are stoloniferous and ephemeral, while their distal regions are rhizomatous and perennate.[98] *Cyperus esculentus* and *C. rotundus* have erect shoots that produce rosettes of leaves in the axils of which buds give rise to typical subterranean stolons with long, slender internodes and vestigial scale-leaves. In *C. esculentus* the apical portions of these stolons may either change their diageotropic growth to an orthotropic one and, thus, give rise to new erect, leafy shoots, or they may metamorphose into thick tubers, consisting of a number of internodes.[99,100] The apical bud terminates the tuber and contains a number of compressed internodes and scale-leaves. The regeneration buds situated on the most proximal (lowermost) nodes of the apical bud are usually those from which new erect leafy shoots eventually develop.[99] In *C. rotundus* the apical portions of the stolons may also change their growth pattern to form erect, leafy shoots, or they may metamorphose into thick tubers, consisting of a number of internodes. They differ, however, from *C. esculentus* in that one or more of the buds (lateral as well as apical) on the tuber may grow into stolons that repeat the pattern, forming chains of tubers. The remaining buds on the tuber may remain dormant for extended periods.[78,101] The stolon of *C. rotundus* also differs from that of *C. esculentus* by its capacity to form lateral buds in the axils of its vestigial scale-leaves, from which branch stolons develop, and by its greater longevity and more protracted role as a carrier of regulatory information (see below). The different role that the tubers play in the reproductive biology of these two related species is reflected in their allocation of dry matter to the tubers: approximately 50% in *C. rotundus* vs. about 29% in *C. esculentus*. Furthermore, while in the former most of the dry matter allocated to tuber formation is shared by a few large tubers, in the latter it is divided between numerous small tubers.[102] Tuber and stolon anatomy of *C. rotundus* and *C. esculentus* have been described by Wills and Briscoe[103] and Wills et al.,[104] respectively.

A tuberization pattern similar to that of *C. esculentus* is exhibited by a number of aquatic species. *Nelumbo nucifera* forms diageotropic stolons in which, at the end of the growing season, the terminal 1-2 internodes form a large tuber loaded with starch. The terminal bud resumes growth as an erect shoot in the following spring.[105,106] *Scirpus grossus*[107] and a number of *Sagittaria* species[92,108] produce numerous typical stolons with long internodes and vestigial leaves. During the growing season the apical bud may change its diageotropic growth into a orthotropic one and give rise to new, erect leafy shoots. Towards the end of the growing season, stolon apices become positively plagiotropic and after growing deeper into the soil they metamorphose into thick, starch-laden tubers, consisting of one to two swollen internodes that subtend the apical bud itself.[109]

A different strategy in the division of roles in diageotropic shoot systems is exhibited by weedy species of the aquatic genus *Potamogeton*. In *P. filiformis* and *P. pectinatus*, lateral buds at the distal end of these shoots do not act merely as regeneration buds, but are also transformed into tubers, in which starch accumulates in the two swollen basal internodes. The apical bud of the tuber sprouts in the following season, forming a leafy shoot.[108,110] In the terrestrial *E. palustre*, lateral buds on the diageotropic, subterranean shoot may develop into erect, green shoots or into short, thick tubers.[111,112] The frequency of such tubers increases as the diageotropic shoot on which they are produced penetrates deeper in the soil.[113] Combination of spreading (stolon) and perennating (rhizome, tuber) strategies is not necessarily limited to the same plagiotropic shoot. Several of the most noxious weeds are able to produce simultaneously plagiotropic shoots that differentiate into stolons, or rhizomes, respectively (e.g., *Cynodon dactylon, Pennisetum clandestinum*). Furthermore, other weeds produce both rhizomes and creeping roots with adventive buds (e.g., *Solidago canadensis, Erigeron caespitosus, Campanula rotundifolia*).[114] Morphological characteristics of spread by creeping roots will be described in Section II.B.2.

Vegetative reproduction is not necessarily dependent on the production of spreading organs. Detachable vegetative propagules are formed instead of flowers in viviparous species (inflorescence bulbils in *A. veneale*)[76] or develop from axillary buds as in *Cardamine bulbifera* and *Ranunculus ficaria*. In *R. ficaria*, the axillary bud produces a short swollen root that functions as the storage organ of the propagule (root bulbils).[115] Some temperate hydrophytes respond to adverse environmental conditions (nutrient depletion, end of season) by producing special detachable vegetative propagules that serve specifically for perennation and propagation in aquatic environments. In several aquatic weeds, such as *Elodea canadensis, Ceratophyllum demersum*, and *Stratiotes aloides*, these structures are not very specialized. They develop from growing shoots and consist of dormant apices enclosed in a dense cluster of reduced, food-storing leaves. Usually they are released by decay of the parent plant. Other species develop specialized dwarf vegetative shoots or buds, known as turions, that have no exact terrestrial counterparts. Turions are characterized by very short internodes and specialized food-storing leaves or scales. In *Hydrocharis morsus-ranae* turions are detached as a result of formation of abscission layers, while in *Hydrophyllum verticillatum* and *Hydrilla verticillata* they are released after decay of the parent plant. Production of these structures is under environmental control (see Section III.A), and their main function is propagation and overwintering. They are protected by mucilage or a thick cuticle, and remain dormant throughout the winter.[108,109,116]

2. Adventitious Buds

Roots and shoots may arise adventitiously from a variety of plant organs and tissues, and this may occur autonomously, i.e., as part of the developmental pattern of the plant, or it may be induced as a result of fragmentation of subterranean organs by external agents, such as cultivation, accidental exposure to unfavorable environment, etc.[117] As was pointed out in Section II.A of this chapter, vegetative reproduction can only take place when the growing regeneration bud is supported by its own functionally associated root system, which most commonly is adventitiously formed from the basal portions of such buds. This distinction eliminates many types of adventitious buds and roots from having a role in vegetative reproduction and, therefore, excludes them also from the scope of this chapter. This applies, for instance, to adventitious buds that develop on cauline organs, such as the hypocotyl, epicotyl, or basal portions of the shoot, that may have a clear-cut role in regeneration (e.g., seasonal, after grazing, cultivation, fires, etc.), but not in vegetative reproduction or even in spread. Adventitious bud formation on leaves frequently contributes to vegetative reproduction. This is a well-known phenomenon in various genera and species of the Crassulaceae, especially in *Bryophyllum* spp., that regularly produce entire adventitious plantlets (carrying leaves and roots) between lobes of their succulent leaves. These plantlets become easily detached by slight mechanical disturbances or by their own weight, and fall to the ground where they become rapidly established. However, this is not a widespread phenomenon and is virtually unknown in weeds. An example of a weed that reproduces by means of plantlets that develop on its leaves is *Cardamine pratensis*.[98]

The major role of adventitious buds in the autonomous vegetative propagation and spread is played by those that are formed on roots ("root buds"). Adventitious buds occur on roots even at considerable depth, as in *Euphorbia esula*[118] and *Asclepias syriaca*,[119] where they presumably contribute to persistence of these weeds under cultivation.

Different root tissues may be involved in the formation of root buds, depending on the developmental stage of the root region in which such buds are initiated.[120] In young roots, adventitious buds are normally produced from the pericycle and the sites of initiation are similar to those of lateral root primordia within the same species (*E. esula*,[121] *Convolvulus arvensis*,[122] *Cirsium arvense*,[123] *Chondrilla juncea*[124]) In older roots, root buds are usually produced after some cambial activity had taken place (*E. esula*,[121,125] *Linaria vulgaris*[126]). In these roots, the initiation of adventitious buds is commonly associated with the emergence of lateral roots.[127,128] These buds may arise from different tissues in the immediate vicinity of the emerging lateral root, including the phellogen (*Chamaenerium augustifolium*,[129] *Armoracia rusticana*[130]) and cortex of the main root (*E. esula*,[131] *Hieracium florentinum*[120]) or in the outer tissues of the lateral root itself, near its point of emergence from the main root (*L. vulgaris*[126]). The physiological basis for this association between adventitious bud initiation and the site of emergence of the lateral root has not been elucidated. It must be remembered, however, that penetration of the external tissues of the main root by the outgrowing lateral root exposes them to mechanical stress, as well as to a variety of substances excreted by Golgi vesicles of the young root cap, including enzymes, and possibly growth regulators.

In some instances root buds play only a minor role in propagation and do not contribute significantly to spread. Among weeds, this is the case in *Rumex acetosa* and *Charefolium silvestre*, which produce adventitious buds at the root crown. In the former, such buds do not produce initially their own root systems. As the plant ages and the root crown thickens, it eventually splits longitudinally and forms separate shoots by partially or entirely separated segments of the thick taproot. In the latter, the adventitious buds do take root and become separate individuals after the older plant completes its life cycle and dies back.[98] In contrast, certain species exhibit a distinctive developmental strategy of vegetative reproduction and invasive spread by means of adventitious buds formed on specialized roots. Some of these are extremely noxious weeds that have proved very difficult to control. *Cirsium arvense*,[132] *Cardaria draba*,[133] *L. vulgaris*,[126] *Centaurea picris*,[134] *Convolvulus arvensis*,[135,136] and *E. esula*[137] exhibit essentially the same features in their developmental pattern of root buds. The plant forms a primary, positively geotropic root, from which permanent lateral roots grow diageotropically for a considerable distance before changing direction to positively geotropic growth. The region at which this developmental change takes place exhibits intensive morphogenetic activity of two kinds. One of these is the formation of a branch root at the bend, that also grows diageotropically and in the same direction as the original root, thus, extending the radial spread of the horizontal roots. The second activity is the formation of several adventitious buds at the bend, or in the region that just precedes it, from which short orthotropic rhizomes develop. These become aerial shoots when they reach the soil surface, but may first branch from lateral buds in the axils of their scale-leaves. The diageotropic lateral roots that branch from the bend also eventually change direction and become negatively geotropic, thus, repeating the pattern again and again. In addition to these creeping root systems that have specialized in invasive spread and vegetative reproduction, short, thin, ephemeral laterals are also formed, usually from the positively geotropic portions of these roots. These have the normal function of roots, namely, the absorption of water and minerals, and play no role in vegetative propagation. Tschirch[138] used the term "heterorhizy" to distinguish between long and short roots in herbaceous plants, that differ not only in the extent of terminal growth, but also in the degree of branching, in the occurrence of secondary growth, in the occurrence of adventitious buds, and in longevity.

In other weed species, there are no indications of an association between the site of formation of root buds and a specific morphogenetic feature of the root on which they are formed. *Apocynum cannabinum*[139] and *Gonolobus laevis*[140] (= *Cynanchum laeve*[141]) produce a system of diageotropic roots that spread radially, without becoming positively geotropic at any time, but which, nevertheless, produce adventitious buds at apparently nonspecific locations along their length. A strategy that combines both types of diageotropic roots is exhibited in *C. arvense*.[142] One type of diageotropic root develops adventitiously from the subterranean portions of the stem or from the primary root, and becomes positively geotropic after growing horizontally a relatively short distance. Adventitious buds, as well as diageotropic lateral roots, develop in the immediate proximity of the bend. The horizontal portions of roots that connect the parent plant to the adventitious plants which develop from these buds are short-lived and eventually disintegrate during winter, allowing the new plants to become independent.[143] The other type of diageotropic roots develop from the primary root and elongate rapidly, without ever becoming positively geotropic. Numerous adventitious buds are produced at apparently nonspecific locations along their entire length, and these, in turn, produce their own adventitious root system. The dual system of diageotropic roots in this species, thus, performs somewhat different roles in radial spread and vegetative propagation. Whereas the former leads to formation of new, independent individuals in relatively close proximity to the parent plant, the latter maintains an interconnected and, therefore, physiologically integrated clone that is mutually supportive (as outlined in Section II.C) and endowed with a marked capacity for rapid invasive spread.[142]

A strikingly different dual system of diageotropic roots has been reported in *Iva axillaris*, a weed native to Canada. A shallow root system is responsible for spread by adventitious bud formation in years with adequate rainfall, while a much deeper lying root system remains virtually inactive. In years with lower rainfall these deeper-lying roots take over the function of spread, by growing upwards and producing regeneration buds close to the soil surface. They then resume positive plagiotropic growth and once more penetrate to deeper soil layers.[142]

C. Intraclonal Communication: Persistent and Nonpersistent Connections

Regeneration buds can be visualized as containers for apical shoot meristem(s) and primordial organs. Such buds are connected to the original plant, as well as interconnected to each other by tissues. Initially these tissues conduct to the developing regeneration buds water and nutrients, as well as assimilates and hormones that regulate their growth, morphogenesis, dormancy (if any), and development. Most of the connecting tissues are mature, in the sense that they exhibit only a limited capacity for continued meristematic activity. The persistence of these connecting tissues is, therefore, primarily a function of their longevity, tolerance of adverse environments, and resistance to attack by biotic factors. However, separation of the individuals from their originators may be part of an inherent developmental process that is basically similar to abscission of leaves, flowers, and fruits (e.g., turions, aerial bulbils, or aerial plantlets formed on leaves of *Bryophyllum* spp.). In these cases, the daughter plants usually become separated at a very early, but well-defined stage in their ontogeny. They usually consist of a single bud, or a group of closely associated buds, that may be protected externally by specialized tissues or modified organs, and include one or more initials of adventitious roots. Such structures are functionally propagules with a far greater natural potential for spatial displacement than other regeneration buds that become separated by unprogrammed means. They appear to have a clearer role in vegetative propagation, but judging from the frequency with which they are encountered in nature, they are by no means a more successful strategy for vegetative propagation, because as a rule they are larger and heavier than the average seed and are not equipped with means that facilitate dispersal as many seeds are.

With the exception of plants whose regeneration buds are programmed to become detached by abscission, regeneration buds of most other plants that spread and reproduce vegetatively remain attached for varying periods. Some of these connections persist relatively briefly and may be considered ephemeral. Others may persist for years, possibly for the entire lifetime of the entire plant system, unless disrupted by external agents.

Quantitative data on the longevity of connecting stems or roots is very scant. In transplanted clones of *Achillea millefolium*, most of the rhizome system was intact and connected to the initial stock 2 years later.[144] In *O. latifolia*, stolons are thin and die during winter resulting in the spread of bulbils.[45] In *Cirsium arvense*, nonperennial parts of the root system rot and disintegrate during the winter, resulting in separation of regions bearing adventitious regeneration buds.[143] In *Rubus fruticosus*, daughter plants are formed at the distal end of trailing shoots and are initially connected to the parent plant. After the daughter plants become established, the trailing shoots senesce, severing the connection to the parent.[29] In other species, daughter plants become disconnected by progressive death and decay of the older parts at the center. This happens in *Pteridium aquilinum*[145] and *Equisetum palustre*,[111] that spread by rhizomes, as well as in *C. avense*[146] and *Hypericum perforatum*[80] that spread by creeping roots. In some species spreading organs are long-lived, as in *Paris quadrifolia* and *Polygonatum verticillatum* (17 years).[147]

Persistence of tissue continuity cannot serve as a sole criterion for functional continuity. The crucial question is what role these tissues play as a system of communication, supply, integration, or correlation between the individual shoot-root systems. If the connection is purely mechanical, it serves only to maintain the position of the individuals relative to each other. To all intents and purposes they are separate individual plants of the same clone, held within the same rigid matrix. On the other hand, as long as the connection retains its vascular functionality, the entire clone remains integrated, allowing the interconnected members to be mutually supportive. Clone partners with access to a more plentiful supply of water, nutrients, or solar energy can translocate the raw materials that they absorb, or the products that they manufacture from this raw material by means of the solar energy that they absorb (photosynthate, photosynthetically derived oxygen, or reduced nitrogenous compounds) to their less fortunately located partners. This is of immense survival value in overcoming competition, as well as in hastening recovery after catastrophes, such as grazing, fire, agrotechnical interference, etc. In *Saxifraga sarmentosa*[148] and *A. repens*,[149] labeled assimilates are translocated from clone partners in the light to those in deep shade through the interconnecting stolons. In *Viola* spp., clone partners can exchange photosynthate through the stolons, and the direction of flow is determined by the relative strength of sinks and sources.[56] The survival value of the mutually supportive interconnected shoot system is evident after defoliation, or partial destruction of one or more of the interconnected shoots, as indicated by the translocation of labeled photosynthate from the intact shoots to the defoliated ones.[51,52,149] Similar situations may occur after partial destruction of spreading weeds by cultivation or herbicide treatment. Similar relationship exist between clone partners developing from adventitious buds on creeping roots of *Sonchus arvensis*.[150]

Correlative phenomena that have been reported as depending on the integrity of clonal interconnections, via internodes or roots, usually involve inhibition of activation of regeneration buds. However, inactivity of a regeneration bud may be due to noncorrelative factors that are either endogenous or are imposed by the external environment. In addition to functioning as channels of communication by which dormancy of buds is controlled, clonal interconnections also serve for the transmission of signals that control morphogenetic activity of the growing bud. Most commonly this is exhibited in the transformation of the developmental pattern from that of a plagiotropically spreading shoot system to that of an orthotropic (erect) leafy shoot. These aspects are dealt with in detail in Sections II.D.1 and II.D.2.

D. Disconnection: Developmental and Physiological Implications

1. Rhizomes

a. Correlative Effects

The effects of disconnection on the different aspects of the vegetative propagation of weeds have been studied mainly in rhizomatous grasses, such as *Agropyron repens, Sorghum halepense, Cynodon dactylon*, and *Panicum repens*. Most of the effects of disconnection were expressed on the dormancy of the axillary buds, but some also involved morphogenetic responses, i.e., on subsequent development of such buds into erect leafy shoots or into diageotropic branch-rhizomes carrying scale-leaves.

One of the most noxious, widespread, and, therefore, most studied weeds is *A. repens*. The plant spreads by means of rhizomes that exhibit a typical apical dominance. Decapitation of the apical bud activates the most distal axillary bud.[151] This apical dominance was strongly reduced when the plants were grown at high nitrogen.[152,153] Detachment of the rhizome from its leafy "parent"increased apical dominance, reversing the effects of high nitrogen.[154] In contrast, Rogan and Smith[155] reported that detachment of the rhizome from its "parent" shoot activates the axillary buds on the rhizome. Thus, while only 3% of the axillary buds on the intact rhizome became activated, either decapitation or detachment caused a considerable increase in activation (to 41 and 54%, respectively) and in combination caused even more activation (to 72%). Several studies were made on the mechanism of these effects. According to Leakey et al.,[156] apical application of an auxin (1-NAA) and/or a cytokinin (6-BAP) to decapitated, attached rhizomes of *A. repens* substituted for the effects of the apex in maintaining dormancy of the axillary buds. However, in decapitated rhizomes that were also detached from their "parent" shoot, it was necessary to apply the growth regulators to both cut surfaces (basal and apical) in order to maintain all the buds along the segment in a state of dormancy. This suggested that both the apical bud and the parent shoot were controlling the dormancy of the lateral buds by means of endogenous growth regulators. However, an exogenous supply of growth regulators to the apical cut surface of the decapitated, attached rhizome failed to inhibit bud activation when the "parent" shoot was at anthesis. This may be interpreted as a developmentally induced change in the sensitivity or receptivity of the dormant axillary buds to the inhibitory effects of these growth regulators. However, in nature, onset of branching of the rhizome coincides with anthesis. Therefore, a simpler explanation might be that by the time of anthesis of the "parent" shoot the axillary buds of the rhizome had already become released from dormancy (possibly as a result of the change in the supply and/or composition of the endogenous growth regulators coming from the "parent" shoot). Consequently, their growth could no longer be inhibited by an exogenous supply of growth regulators that act specifically in maintaining an existing state of dormancy.

In *A. repens*,[155] *Poa pratensis*,[157] and *C. dactylon*.[158] detachment of the rhizome from its "parent" shoot resulted in activation of the dormant regeneration buds, while the apical bud changed its growth from diageotropic to orthotropic and was transformed into a leafy shoot. In *A. repens*, application of GA to the cut surface at the basal cut of the rhizome reversed the activation of the axillary buds, as well as the transformation of the apical bud.[155] While it is possible that an acropetal flux of GA controls both these processes independently, it seems more likely (on the basis of data to be discussed below) that parental supply of GA controls morphogenesis of the apical bud, which, in turn, releases the correlative control that it exerts over the activation of the axillary buds. The auhors also showed that while IAA failed to similarly inhibit bud activation when supplied through the apical cut surface of decapitated rhizomes that remained attached to their "parent" shoot, it maintained a high level of inhibition in the axillary buds when supplied to the basal cut surface of detached, nondecapitated rhizomes. The possibility that these differences in response to IAA when supplied to the basal or apical cut surfaces, respectively, could be a result of polar transport of the IAA that was not studied. Application of kinetin to the subapical axillary bud of rhizome segments of *A. repens* released it from dominance by the apical bud.[159] as has been reported for apical dominance in leafy shoots.[160] Supply of nitrate and sucrose through the cut basal end of a detached, nondecapitated rhizome of *A. repens* decreased the level of dormancy in the axillary buds, indicating that "parental" control was not necessarily mediated only by hormones, but may involve nutritional factors. These nutritional aspects were studied by McIntyre. He found[153] that as the level of nitrogen supplied to the plant was increased, the rhizomes branched more profusely. At all levels of nitrogen, decapitation of the rhizome activated the subapical axillary bud and activation was accelerated by increasing the levels of nitrogen.[159] Even in detached, nondecapitated rhizomes, the subapical axillary bud was activated when sucrose was fed through the cut end, as well as when the detached rhizome was long.[154] The author concluded that that inhibiton of the subapical bud in detached, nondecapitated rhizomes is due to its inability to compete effectively with the apical bud for the limited supply of carbohydrates. Subsequent development of buds activated by rhizome decapitation depends on the environmental conditions under which the "parent"plant grows. In *A. repens* growing at high temperature, decapitation of a rhizome that remained attached to the "parent"shoot activated the more distal axillary buds to grow into orthotropic leafy shoots, while the less distal buds were activated to grow diageotropically as rhizomes. However, a lower temperature inhibited the geotropic transformation of the activated distal buds and caused them to grow as branch rhizomes. This inhibition was reversed when the "parent"shoot growing at the low temperature was exposed to a lowered light intensity, upon which the rhizomes produced more leafy shoots from their axillary buds.[151] The tendency of rhizome buds to develop into leafy shoots was ascribed to reduced supply of assimilates from the "parent" shoot growing at low light intensity and high temperature.

b. Positional Effects

Positional differences in the response of the axillary buds to decapitation of an attached rhizome were observed in *A. repens*.[151] Buds at some distance from the apex were activated to form branch rhizomes, irrespective of whether the rhizome was decapitated subapically or closer to them. Their morphogenesis, thus, appears not to be a result of apical dominance. Buds in a more basal position appeared to be inherently dormant, inasmuch as they could not be activated even when decapitation was performed at a node close to them. This dormancy was clearly not due to apical dominance. On the other hand, when the "parent" shoot was exposed to reduced light intensity, the inherent dormancy extended to more distal buds. This suggested that the level of assimilates may be limiting the capacity of the more basal buds on the rhizome from becoming activated in response to removal of apical dominance.

Positional differences in activation potential of the axillary buds along the rhizome do not apparently depend on attachment to the "parent" shoots. Acropetal gradients were observed in detached and decapitated rhizomes of *A. repens*,[155,156,161,162] *S. halepense*,[163,164] *C. dactylon*, and *P. repens*.[165] Fragmentation of the rhizome into multinodal segments showed that positional differences in the potential of the buds to become activated and, subsequently, grow were maintained along the segment, irrespective of its size (*A. repens*[155]) or of the region of the rhizome from which it was excised (*S. halepense*,[163] *A. repens*,[155] or *C. dactylon*[165,166]). The positional gradient in the potential of axillary buds to become activated, as well as in their subsequent growth rate, was maintained even in single-node segments of the rhizome (*S. halepense*,[163] *A. repens*,[167] *C. dactylon* collected in spring[168] or *P. repens*[166]). It, therefore, appears that this gradient is not a direct consequence of the establishment of a dominant shoot from a bud close to the distal cut of the segment. On the other hand, since these inherently dormant buds can, nevertheless, be activated, it is unlikely that their lack of response to decapitation was merely due to substrate deficiency, as suggested by McIntyre.[151] These studies, with detached and decapitated rhizomes and with rhizomes fragmented into multinodal and single-node segments, do not report on the subsequent morphogenesis of the activated buds. It is, therefore, unknown whether or not the positional differences in morphogenesis are affected by the number of nodes in the segments, their proximity to the apex of the rhizome, or their proximity to the distal, cut surface. Such information is available from detached, nondecapitated rhizomes of *P. pratensis*.[157] In such rhizomes, the apical bud transformed into an erect leafy shoot. Only the axillary buds on the seven nodes subtending the apex became activated, and of these, the more proximal to the apex grew into branch rhizomes while the less proximal ones grew into erect, leafy shoots.

Despite the indications to the inherent nature of the positional gradient in the potential of the axillary buds to become activated, as well as in their subsequent growth rate, the existence of such a gradient was strictly seasonal in *A. repens*[167,169] and *S. halepense*[163] and its direction was seasonally reversed in *C. dactylon*.[168] The possibility that imposition of this gradient was related to the distribution of nutrients is, therefore, of interest.

The acropetal gradient in the activation of axillary buds on detached[161] and on detached and decapitated[170] rhizomes of *A. repens* was associated with an acropetally increasing gradient in nitrogen content. Moreover, increasing the supply of nitrogen extended bud activation basipetally and also enhanced the transformation of the more basal-activated buds into erect leafy shoots, rather than into diageotropic rhizome branches with cataphylls. Leakey et al.[171] reported the absence of a gradient in soluble sugars, starch, nitrate, and nitrite content in the axillary buds along the rhizome of *A. repens*, but the concentration of ammonia increased progressively in an acropetal succession. According to Nigam and McIntyre,[170] the acropetal gradient in the potential of bud activation in *A. repens* was positively correlated with the amino acid content of the buds and rhizome nodes, especially with asparagine and glutamine levels. In rhizomes from plants grown at high nitrogen levels, the gradient in amino acid content was considerably reduced in parallel to the more uniform activation potential of the buds along the rhizome. On the other hand, Moreira and Rosa[168] found no positional correlation between the content of sugars or starch in the axillary buds of *C. dactylon* and their growth potential. The possibility that the positional effects on the activation potential, subsequent growth rate, and morphogenesis of the axillary buds might be correlated with endogenous levels of growth regulators does not appear to have been studied.

c. Correlative Inhibition within Segments

Fragmentation of a rhizome into multinodal segments often induces activation in the majority of the buds. The proportion of activated buds increases as their number per segment is less (shorter segments). This negative correlation between number of buds (segment length) and the proportion of sprouting buds has been reported in *A. repens*,[208] *Achillea millefolium*,[172] and *S. halepense*,[173] and most probably indicates correlative effects among the associated buds. Moreover, an acropetal gradient in growth rate of the activated axillary buds was soon established along the rhizomes segment of *Agropyron repens*.[155,156,174] A similar response has been reported in *S. halepense*,[163,164] *Digitaria scalarum*,[175] *C. dactylon*, and *P. repens*.[165,166] While the fastest growing bud(s) continued to grow, the remaining buds exhibited a progressively diminishing growth rate the farther they were from these buds. Growth of these temporarily activated buds was arrested for long periods, but they retained a latent capacity for resuming growth, e.g., when isolated as single-node segments[176] or when the dominant shoot had been excised.[177] In the latter case, one of the reactivated buds assumes dominance. Similar results were reported in *C. dactylon*.[165]

Time-course studies of axillary bud activation in multinodal segments of *A. repens*[174] indicated a polarity in the establishment of dominance, which was most frequent in the subterminal bud. However, dominance was not necessarily associated with the bud that exhibited the fastest initial growth rate. In binodal segments of *S. halepense* rhizomes, both axillary buds initially became activated, but one of these became dominant quite early, suppressing growth of its companion and eventually killing it. This dominance was equally shared by the distal and proximal buds.[178] Activation and continued growth of adventitious roots from the nodes of rhizome segments of *A. repens* were not necessarily correlated with growth or even survival of the associated bud.[174]

d. Effects of Spatial Orientation

Spatial orientation of multinodal rhizome segments did not affect the polar character of dominance in activation and growth of axillary buds in several weed species belonging to the Gramineae.[163,165,179] However, in nondecapitated multinodal segments of *C. dactylon* and *P. repens*, a downward orientation inhibited growth of the apical bud, and this apparently caused partial release of the axillary buds from dormancy.[165] In *Cyperus rotundus*, each tuber consists of a number of internodes, and when it was isolated, activation of its axillary buds was strictly polar: the more distal bud on the intact tuber, as well as on the bottom half of a tuber sliced equatorially, was the first (and usually the only) one to sprout.[180] In the related species *C. esculentum*, the related oldest, most basal bud to differentiate on the tuber is situated on the thickest node and is also the largest and the first to sprout on the intact tuber. When disconnected from the tuber chain, the remaining buds sprout in acropetal succession, but in the nondisconnected tuber, sprouting ceases after the second bud.[99] In *C. rotundus* the polarity of dominance was not at all affected by the spatial orientation of the individual tuber. In contrast, multituber segments of the subterranean shoot system exhibited dominance of the apical tuber only when the chain was planted upright, not when it was planted horizontally or upside down (all tubers sprouted). Exposure of the intertuber portions of the shoot system to hot water or to alcohol eliminated all dominance and resulted in sprouting of all tubers. Nevertheless, even when the tuber chain was intact, its exposure to light induced bud activation in all tubers.[180] The modification of polar dominance in detached segments of tuber chains of *C. rotundus* by their spatial orientation is similar to that observed in apical dominance of intact erect aerial shoots that had been trained to grow horizontally.[181]

e. Environmental Effects

Nitrogen enrichment prior to detachment modifies the apical dominance in axillary bud activation along nondecapitated rhizomes of *A. repens*.[154] However, nitrogen enrichment after fragmentation failed to influence polar dominance in bud growth on multinodal rhizome segments of *A. repens*[176,182] and *C. dactylon*.[165] On the other hand, exposure of the fragments to high temperature reduced or considerably delayed the establishment of dominance in *A. repens*,[182] *C. dactylon*, and *P. repens*.[165]

Light inhibited the onset of axillary bud activation in multinodal rhizome segments of *D. scalarum*[175] and *A. repens*.[183] However, when light was applied after the buds had sprouted in darkness, it prevented the onset of dominance, and all buds grew equally (loc. cit.). This explains why shallowly buried segments of *A. repens* rhizomes can produce several shoots, and if these emerge before dominance is established they are no longer subject to mutual inhibition.[183] Elimination of the shorter wavelengths from (incandescent) light reduced the initial inhibitory effects of light on bud activation in *A. repens* rhizomes, but the data are insufficient to support the conclusion that this was due to spectral effects[183] and not to reduced irradiance.

2. Roots

a. Correlative Effects

Adventitious buds are present on roots of many species, some of which are extremely noxious weeds. It is, therefore, surprising that effects of detachment on activation and morphogenesis of root buds has not received as much attention as in rhizomes. In a number of *Rubus* species,[184,185] as well as in *Populus tremula*,[186] removal of the aerial shoots (or even of their apices) enhanced sprouting of "suckers" from root buds. This effect was reversed by exogenous auxin. However, in *Euphorbia esula* moisture content of the roots (and presumably their water potential) increased significantly after detopping, and it has been suggested that sprouting of such buds was inhibited by competition for water by the shoot (or by water stress).[188]

In seedlings of *E. esula*, removal of aerial shoots accelerated sprouting of preformed root buds within 24 to 48 h, some of which continued to grow, while others were reinhibited.[188-190] This correlative inhibition between neighboring root buds was reduced by increasing levels of nitrogen during growth of *Cirsium arvense*[187,191] and *E. esula*.[192] In *Chondrilla juncea*, adventitious buds are initiated in the root only after it had been separated from the shoot. Studies with exogenous growth regulators suggest that initiation of adventitious buds may be inhibited in intact attached roots by endogenous auxin originating in the shoots.[193] In *C. arvense* the number of adventitious buds produced per unit length of root segment was inversely related to the length of the segment.[194] These results indicate the existence of correlative inhibition by the shoot on initiation of adventitious buds and on sprouting of preexisting adventitious buds, as well as imposition of inhibition by some of the sprouted buds over the continued growth of their partners. Analogous promotive effects of disconnection and correlative inhibition in their sprouting and subsequent growth of interconnected axillary buds on rhizomes were described above. Therefore, the mechanisms of correlative inhibition by the leafy "parent" shoot, as well as by the adjacent buds that achieve dominance, are basically the same in axillary and in adventitious regeneration buds.

b. Polarity

Fragmentation of roots of *C. juncea* induced initiation of adventitious buds on the proximal portion of the segment and of roots on its distal portion, regardless of the length of the segment (2 to 20 cm) or of its location along the root.[193] A similar polarity in the morphogenetic capacity of root segments has been found in other weeds, such as *Taraxacum officinale*,[195] *Chamaenerium augustifolium*,[129] *Convolvulus arvensis*,[196] *Cirsium arvense*,[194] and *Rubus procerus*.[197]

The inherent polarity in the initiation of adventitious buds and roots along root segments has been attributed to a differential distribution of endogenous hormones in the segment. In general, increasing levels of exogenous auxin enhance root initiation, but inhibit adventitious bud initiation on root segments in vitro. On the other hand, cytokinins usually enhance adventitious bud production.[120] However, in *C. arvensis*[196] and *C. juncea*,[193] lower concentrations of auxin stimulated bud initiation. Thus, it has been suggested[193] that the morphogenetic polarity in initiation of adventitious buds and roots in root segments may be associated with a change in auxin-to-cytokinin ratio, resulting from removal of the source of cytokinins at the distal end and of the source of auxin at the proximal end, as well as to the polar transport of auxin along the root.

c. Critical Size

The size of the root segment affects its regenerative capacity. In *Armoracia rusticana* the segment must include a root scar and its associated root trace for regeneration to occur.[198] In *T. officinale* and *Cichorium intybus*[195] and in *Cirsium arvense*[194,199] a lower limit was found below which the size of the root segment becomes limiting for regeneration. In *Convolvulus arvensis*, exogenous auxin can restore the regenerative capacity to root fragments of limiting length.[196] Thus, critical size may depend on availability of phytohormones, or other substances that are essential for the morphogenetic process.

d. Environmental Effects

Most studies on environmental effects (light, temperature, nutrients, etc.) on regeneration from roots are based on the number of emerged buds or shoots after exposure to the factor being studied. It is, therefore, difficult to evaluate to what extent these effects are on initiation of adventitious buds, on activation of new and preexisting buds, or merely on bud growth.

According to Peterson,[120] light has little or no direct effects on bud initiation in root segments, but strongly affects their subsequent development. Consequently, the fact that adventitious buds on root segments of *C. arvensis* make no further growth unless they are exposed to light, but at the same time light depresses bud initiation,[200] may be attributed to the establishment of dominance by the bud(s) whose growth is first light stimulated. The response of the bud to light was wavelength dependent, suggesting regulation by the phytochrome pigment system. Although buds were initiated very soon after fragmentation, only older buds were responsive to (red) light.[201] The results suggest that, in the field, root buds may be initiated, but remain inactive unless the roots grow sufficiently close to the soil surface to perceive light or are exposed to light by cultivation.

3. Ecological Consequences and Implications of Fragmentation

Fragmentation of subterranean organs occurs as part of cultivation of the soil, which also results in burial of the segments over a range of depths in the soil. The more intensive the cultivation, the greater is the frequency of smaller fragments.[97] The potential of fragments to regenerate and establish plants, as a function of their size and depth of burial, was studied in roots that form adventitious buds (*C. juncea*,[202] *C. arvense*,[194] *Sonchus arvense*,[203] *T. officinale*[204]), as well as in rhizomes that carry axillary buds (*A. repens*,[205-208] *C. dactylon* and *P. repens*,[209] *Pennisetum macrourum*[210]). The results were similar, namely, the probability for successful emergence and establishment decreased with depth of burial, but increased with size of the segment. Furthermore, in *A. repens*, the frequency of spike-bearing culms increases in plants regenerating from larger rhizome segments.[205] Depth of burial had no effect on the sprouting capacity of two-node rhizome segments of a number of species, irrespective of whether or not they had eventually succeeded to emerge.[211] It is, therefore, likely that larger segments are able to sustain for longer periods heterotrophic growth of the bud that had established dominance over its partners. In this may lie the adaptive value of dominance among partner regeneration buds, since it increases the probability of emergence and establishment of a new shoot. Prolonged periods of heterotrophic growth until emergence may explain the reduced vigor of the shoots that emerge from greater soil depths.[205,207,212,213]

4. Conclusions

The picture that emerges from the data and hypotheses summarized above is that an inactive axillary bud on a rhizome or an adventitious bud on a root may exhibit marked responses when it is disconnected, either distally or proximally, or both. The responses range from activation through growth rate to morphogenesis (i.e., transformation of the bud into an orthotropic leafy shoot vs. continued growth as a diageotropic branch rhizome with scale-leaves). Disconnection releases the bud from influence by the part(s) that had been removed. Responses to separation of the bud from more apical portions of the rhizome have generally been ascribed to either of two effects. One is the removal of the source of endogenous growth-regulating activity situated in the apical bud itself, or in the rapidly growing apical region of the rhizome. The other is the removal of the strong metabolic sink situated in the rapidly growing apical portion of the rhizome, which competes vigorously with the axillary buds. Similarly, separation from the "parent" shoot stops the import of readily available assimilates from it and immediately makes all metabolic and growth activity in the rhizome entirely dependent on storage. This may involve major changes in biochemical pathways. The same is true for growth regulators whose source is in the "parent" shoot and in its root system. Disconnection also reduces or even eliminates the longitudinal gradients in chemical potential of metabolites and in water potential. All these changes have to be taken into consideration in studies of the mechanism of the effects of disconnection on bud activation, growth, and morphogenesis.

The response of the individual bud may also be strongly modified by the length of the segment on which it is situated. Evidently, the longer the segment, the greater are the possibilities for interaction between buds situated on the same segment. Buds on a segment of a rhizome or root may resemble each other in morphology, but those that are more distal are younger (not necessarily so in roots), while those that are more proximal are probably more juvenile. Superimposed on these gradients is the natural diversity that exists in any normally distributed population. These differences may play a major, albeit a hitherto entirely unrecognized, role in determining which bud would become activated first, what its growth rate would be, and what developmental direction it would take morphogenetically. Whatever course its growth and development would take may have profound effects on the fate of the other buds in the segment. These may range from simple competition for the limited supply of stored food and of growth regulators, through correlative inhibition, to outright killing.

Disconnection involves massive damage to cells at the cut surface(s), in addition to drastic changes in their water relations, as well as in their atmospheric environment. Cell damage may result in release of potent metabolites, as well as in production of "wound hormone". Rhizomes are characteristically thick and fleshy and it is conceivable that the intercellular atmospheric composition changes significantly when the rhizome is cut, inside the soil or in the laboratory. The possibility that the complex changes that most probably take place at the cut surface may have a controlling effect on the neighboring buds has not received adequate attention.

III. DEVELOPMENTAL AND ENVIRONMENTAL ASPECTS OF VEGETATIVE REPRODUCTION

A. Initiation Phase of Spreading and Perennating Organs

1. Introduction

Spread, with or without subsequent vegetative reproduction, is associated with roots and shoots that grow diageotropically or plagiotropically. Initiation of spread is, therefore, concerned with initiation of such organs from other organs that do not contribute to spread, because their geotropic growth is either negative (shoots) or positive (roots). Initiation of shoots and roots that play a major role in spread (and in subsequent vegetative reproduction) may be regulated in different ways. There are few reports of studies of factors that may be directly involved in initiation of such organs. Nevertheless, it appears that in some species initiation is under endogenous control, i.e., part of an inherent genetic developmental pattern that can be modified quantitatively, but not changed, by environment and/or nutrition. In other species, initiation is seasonal, indicating either environmental control or a consequence of other environmentally induced developmental events, such as flowering, dormancy, etc. Finally, age or size of the plant may also determine its capacity to initiate these organs of spread.

2. Rhizomes, Runners, and Stolons

In *Cynodon dactylon* growing under controlled conditions, initiation of runners and rhizomes took place in leaf axils of basal internodes of third-order tillers, and was attributed to ethylene production enhanced by crowding of tillers at the crown.[214]

In other weeds, initiation of spreading organs is seasonal. In *Ranunculus repens*, stolons tend to appear during late spring and summer.[215] In *Solidago canadensis*[216] and *Tussilago farfara*[217] autumn is the main period for rhizome production (after flowering and seed set). In *Rubus fruticosus* the transformation of vertical shoots into trailing plagiotropic ones occurs in late summer.[29,95]

Seasonality in the initiation of spread is usually attributed to some unidentified environmental factor, or combination, that also changes in parallel. In *Fragaria* spp. axillary buds of the rosette develop into stolons under long days, whereas under short days they develop into inflorescences.[218] In *Polygonum thunbergi*, on the other hand, short days enhanced flowering as well as stolon formation.[219] However, surprisingly little is known about specific seasonal factors controlling initiation and growth of spreading organs in weeds.

Initiation of spread by rhizomes and stolons may take place in the earliest stages of development of the erect shoot system. Studies with *Agropyron repens*[220,221] and *Sorghum halepense*[178] grown from seed show that the buds situated at the lowest nodes (down to the coleoptilar node in *A. repens*) sprout almost immediately after seedling establishment and grow directly into rhizomes. Usually there is an acropetal progression in the proportion of axillary buds that develop into tillers rather than into rhizomes. The pattern was virtually the same in shoots of *A. repens* developing from regeneration buds on rhizome fragments[222] or from emerging rhizomes.[70] In seedlings of *A. repens* the steepness of this gradient was modified by environmental and nutritional factors. Lower light energy (shade or shorter photoperiods) and higher temperatures and nitrogen supply increased the proportion of basal axillary buds developing into rhizomes, but the coleoptilar bud was less labile and almost exclusively developed into a rhizome.[220,221] In *S. halepense*, on the other hand, the gradient was not significantly affected by photoperiod, level of solar irradiation, temperature, or by thermoperiod.[178]

Developmental aspects of stolon initiation in weeds has been studied to a much lesser extent. In nonweedy species, such as *Solanum tuberosum*[223] and *S. andigena*,[224] stolons develop from the more basal axillary buds, the first stolon being initiated very soon after sprouting of the "parent" tuber. Thus, it is conceivable that in some stoloniferous and rhizomatous species the developmental behavior of the basal axillary buds of the erect shoot is predetermined to take place during the early stages of seedling development, or after the transformation of the spreading organ into an erect leafy shoot, and environmental factors only modulate this morphogenetic gradient.

In most rhizomatous species, erect leafy shoots are formed by transformation of the apical bud of the rhizome to negatively geotropic growth (*S. halepense*,[225] *A. repens*,[70] *Imperata cylindrica*,[226] *Phragmites communis*,[227] *C. dactylon*[68]). In such transformed shoots, rhizome buds are initiated on the nodes most closely associated with the region where transformation to the erect habit takes place. All the species listed above are monocots, and the apex of the transformed shoot is almost invariably determinate, i.e., terminates vegetative activity, usually by flowering.[228] It is, therefore, possible that initiation of spread may be physiologically associated with either the transformation into an erect leafy shoot or with termination of vegetative activity by the apex. In species that spread by adventitious buds on specialized diageotropic roots, such buds are frequently associated with the region where the creeping root transforms to positive geotropism and to "normal" root function (absorption).[126,133,134,136]

Seedling shoots and transformed shoots have in common the emergence of their apices from the darkness inside the soil to light in the aerial environment. Transitions from darkness to light are frequently signals for morphogenesis. It seems that the role of light in the localization of rhizome initiation to the basal nodes of emerging erect shoots of seedlings and transformed rhizomes has not received adequate study. However, direct exposure to light of the basal portion of the seedling shoot of *Sorghum halepense*[178] or burial or exposure of the basal parts of erect shoots of *A. repens*[70] did not affect the acropetal progression from rhizome buds to tiller buds.

Stolon and rhizome initiation may be under hormonal control. In *Solanum andigena*, decapitation of an aerial shoot induces activation of buds at the upper nodes, which then develop into leafy orthotropic branches. This response is slightly inhibited by applying IAA to the decapitated stump, whereas GA alone increased elongation of the axillary shoot. Simultaneous application of IAA and GA caused the axillary bud to grow as a plagiotropic, stolon-like shoot, with elongated internodes and greatly reduced leaves. If cytokinin was applied to the apex of these stolons or to stolons produced normally, they turned upwards and showed normal leaf expansion.[224,229] The exogenous cytokinin accumulates in the stolon tips prior to their conversion to orthotropic growth.[230] In strawberries, stolon production under long days is also hormonally regulated. Stolon initiation may be induced under short days by gibberellin applications.[231] Thus, changes in the endogenous balance between different plant hormones may play a central role in the production of spreading shoots.

3. Tubers

Tuberization was enhanced by decreasing daylength (short days) in *Cyperus rotundus*,[232,233] *C. esculentus*,[59,234,235] *C. serotinus*,[236] and *Sagittaria trifolia*.[237] There are, however, reports of increased tuberization in *C. rotundus* under long-day conditions.[238,239] This may be due to the increased availability of light energy, since decreased light intensity reduced tuberization and tuber size in *C. rotundus* and *C. esculentus*.[240,241] Tuber size in *C. esculentus* was not affected by photoperiod.[59]

High nitrogen levels reduced tuberization, but promoted the formation of stolons and aerial shoots in *C. rotundus*[242] and *C. esculentus*.[234] Thus, when mineral nutrition is abundant, vegetative spread and establishment of new daughter plants predominates over the production of perennating organs.

4. Turions

Environmental control of turion formation is much clearer and involves an interaction between temperature and photoperiod. In *Myriophyllum verticillatum*,[243] *Hydrocharis morsus-ranae*, and *Spirodela polyrhiza*,[244,245] turions are induced by short days, but the duration of the critical daylength increases with temperature. In *Potamogeton crispus*, turion initiation requires long days and temperatures higher than 12° C. Below 24° C, daylength must be longer than 12 h, but at 30° to 35° C, turions are formed also in 8-h photoperiods.[246]

Nutrient deficiency, especially of nitrogen, induced turion formation in *S. polyrhiza*.[247] It is possible that these effects of nitrogen are indirect, and that when nitrogen is abundant, vegetative growth predominates and inhibits turion formation. The environmental control over the induction of turions is probably mediated by hormones. In *S. polyrhiza*, ABA inhibited growth and induced turion formation.[248] In *M. verticillatum* ABA enhanced turion formation under photothermally inductive conditions, while cytokinins and gibberellin were inhibitory.[243]

5. Other Organs

Practically no information is available on the regulation of initiation of creeping roots. Similarly, although the formation of bulbs has been studied in great detail in economically important species, such as *Allium cepa* and *A. sativum*, as well as in ornamentals, similar studies with weedy species are notable by their scarcity.

B. Growth of Spreading and Perennating Organs

Environmental factors that increase overall plant growth (e.g., soil moisture, light intensity, temperature, nutrient levels) may have parallel effects on the production and growth of spreading and perennating organs. In *I. cylindrica*, for instance an increase in temperature (from 23°/17° to 29°/23° C) or in photoperiod (from 12 to 16 h at constant light intensity) resulted in larger plants, but the relative allocation of biomass to rhizome growth was practically unaffected.[249] Similarly, an increase in photoperiod (at constant intensity) enhanced biomass production in *C. rotundus*, without affecting its partitioning to shoots, roots, tubers, and stolons.[239] However, in other cases, overall growth and the growth of spreading and perennating organs are affected differentially by environmental factors. In rhizomatous weeds, a decrease in the daily input of light energy (intensity or duration) reduced the relative allocation biomass to rhizome growth in *Poa pratensis*,[250] *T. farfara*,[217] *Agropyron repens*,[251] and *I. cylindrica*.[252] Similarly, in *C. rotundus*[241,253,254] and *C. esculentus*,[241,253] a decrease in light intensity reduced total growth as well as the relative amount of biomass allocated for tuberization (fewer and smaller tubers). However, these related species differ in their developmental response to lower light intensity. In *C. esculentus* leaf growth per shoot and the number of tubers produced per shoot were reduced by lower light intensities, while in *C. rotundus* the number of shoots was reduced, but leaf growth and number of tubers per shoot were unaffected.

Light intensity effects are frequently extrapolated and compared to shade effects in the field. However, shade under a crop canopy is also characterized by changes in spectral quality of the transmitted light. These changes in spectral energy distribution have profound developmental effects[255,256] in addition to those on photosynthesis. However, there are, as yet, no published reports on effects that such changes may have on growth of spreading organs, of weedy and nonweedy species, nor on their initiation (see previous section).

Mineral nutrition affects biomass partitioning to spreading organs. In *A. repens*,[153,257,258] *Bromus inermis*,[259] and *P. pratensis*,[260] higher levels of nitrogen increased tillering more than rhizome production. In *T. farfara*, a tenfold reduction in levels of nitrogen or potassium reduced the number of rhizomes produced, without affecting overall growth.[217] Limiting levels of water and nitrogen in the soil affect differentially the elongation and branching of spreading organs. In *A. repens*[261] and *Ranunculus repens*,[262] water stress or low N levels inhibit branching much more than rhizome or stolon elongation, respectively.

The variety of responses to different environmental factors that affect spread and vegetative reproduction in different weed species may reflect different developmental strategies, while conflicting and contradictory results with the same species may reflect differences in the clonal material or experimental conditions used by the different investigators. Furthermore, the environmental factors that affect the production of spreading organs may have different effects on the initiation and on the subsequent growth of these organs. However, most studies fail to discriminate between these two components that contribute jointly to spread and vegetative reproduction.

C. Transition from Spread to Reproduction and Establishment of New Individuals

New individual plants arise asexually when buds (axillary, apical, or adventitious) grow into autonomous, adventitiously rooted erect leafy shoots, which sooner or later become separated from each other. In some cases, the buds sprout and grow into the erect shoots directly, i.e., without undergoing a period of inactivity. This is prevalent in species that spread and reproduce asexually by means of runners, such as *Cynodon dactylon, R. repens, Oxalis corniculata*, and *Commelina benghalensis*, and is especially effective in certain very noxious annual weeds, such as *Digitaria sanguinalis*. In other cases, the buds undergo a period of inactivity. This may arise from correlative inhibition, as evidenced by activation of the bud as a result of its isolation from the correlative effects of the growing apex, of active neighboring buds, or of the "parent" shoot (Section II.D). Correlative inhibition is apparently the predominant factor that regulates the activation of rhizome buds on *Polygonum amphibium*,[263] *Cynodon dactylon*,[68] and *Sorghum halepense*,[164,264] since no seasonal, or inherent dormancy has been observed in single-node segments of these species. Where correlative inhibition exists, disconnection results in bud activation. For weeds, disconnection is a natural consequence of cultivation.

Inactivity of the buds may also be due to inherent dormancy, from which they are released either by environmental signals,[265] or through internal changes. In addition, the transition may take place as a result of transformation of the tropic response of apices of subterranean shoots to negative gravitropism, which results in emergence of the apical bud from the soil to produce an erect shoot. Such transformation may take place while the shoot apex is still within the soil, but in weeds, disturbance of the soil through cultivation may promote this transformation, either by exposure to light of the subterranean apex[158,159] or by morphogenetic consequences of disconnection. These aspects are summarized in Section II.D.1.

In regions with climatically distinct seasons, the vegetative production of new individual plants is frequently seasonal. This may be directly due to the alternation between seasons that are generally favorable for growth with others that are not. On the other hand, the effects may arise indirectly, for instance, by association with other developmental processes that are themselves under seasonal control. Thus, axillary buds on rhizomes (*A. repens*,[266] *P. coccineum*,[267] *Vernonia baldwinii*[268,269]), as well as adventitious buds on roots (*Armoracia rusticana*,[270] *Euphorbia esula*,[190,271] *Chondrilla juncea*,[67,272] *Taraxacum officinale*,[204] *Sonchus arvensis*,[273] *Cirsium arvense*[65]), exhibit seasonal fluctuations in their capacity to become activated and sprout. This capacity is minimal in late spring and early summer, coincidentally with flowering and seed set. Seasonality was also exhibited in isolated segments of the rhizomes and roots, respectively, and is, therefore, not related to correlative inhibition.

In several weeds that spread by adventitious buds on creeping roots (*E. esula, S. arvensis*, and *C. arvense*;[274] *Lychnis alba*[275]) or by rhizomes (*V. baldwinii*[276]), carbohydrate reserves reach their lowest level during flowering and seed set, at which time emergence of adventitious shoots in the field is also minimal. In *A. repens*[169,171,257] and *P. coccineum*,[267] total nitrogen level in the rhizome was lowest during the flowering phase. These results suggest that flowering and seed set may deplete roots and rhizomes of their nutrient reserves, thereby reducing their sprouting capacity. This suggestion is supported by the fact that seasonality was greatly reduced or even absent in buds on single-node rhizome segments of *A. repens* collected from well-fertilized plants.[169,171] Furthermore, sprouting of buds on single-node segments harvested at that time was enhanced in the presence of nitrogenous compounds as well as in presence of gibberellin and partially enhanced by prechilling.[167] The fact that gibberellin and prechilling enhanced sprouting, even in absence of nitrogen supplement, and *vice versa,* places some doubt on the causal relationship suggested between nutrient depletion at time of flowering and the reduced sprouting capacity. In this connection, it is surprising that the association between flowering, nutrient levels, hormone levels, and sprouting capacity has not been studied by artificially suppressing flowering, flower expression, or seed set. If, indeed, changes in sprouting potential are causally associated with flowering, such association will be masked in regions or in species in which flowering is not seasonal. However, in *Sorghum halepense*[164,264] and *Cynodon dactylon,*[69] flowering and rhizome growth occur simultaneously. In *S. halepense* most rhizome growth takes place between flowering and seed set, and photosynthate translocation to rhizomes gradually increases during this phase.[164] Thus, in *S. halepense* seed production has a lower priority for assimilate partitioning compared to rhizome growth.

Cyperus rotundus and *C. esculentus* exhibit seasonal variations in the capacity of sprouting from isolated tubers (*C. rotundus,*[277] *C. esculentus*[278-280]). However, other investigators[281] report the absence of such seasonal variations, since isolated tubers maintained virtually the same capacity for sprouting when harvested at intervals during the year. In those biotypes that exhibit seasonality, sprouting capacity may be substantially increased during the low season by prechilling. This reponse is prevalent in overwintering structures of aquatic plants (*Hydrilla verticillata,*[282] *M. verticillatum,*[284] *Spirodela polyrhiza,*[247] *Hydrocharis morsus-ranae*[285]). In *C. rotundus* the requirement for prechilling differs quantitatively in different geographic biotypes.[233] In *C. esculentus* prechilling increases the number of sprouting buds per tuber, as well as the proportion of sprouting tubers.[279] Similar phenomena have been reported in bulbs of *O. latifolia* and *O. pes caprae,*[286] rhizome segments of *V. baldwinii,*[260] as well as in adventitious buds on roots of *Sonchus arvensis*. Sprouting in the field from roots of *S. arvensis* ceases in early autumn and this was paralleled in root segments under laboratory conditions.[287,288] Sprouting capacity from such root segments was regained after prechilling.[203]

Leaching increased the sprouting capacity in isolated tubers of *C. rotundus*[157,281,289] and *C. esculentus,*[280] indicating the mediation of growth inhibitors. This is supported by studies showing negative correlation between endogenous levels of abscisic acid (ABA) and sprouting capacity of tubers of *C. rotundus.*[290] Also, exogenous growth regulators modify the sprouting capacity of such tubers: ABA reduces it, while cytokinins and ethylene increase it.[277,291,292] Thus, bud dormancy and, in particular, its release by chilling may involve endogenous growth regulators and changes in the balance between them. This has been demonstrated in the aquatic plants *M. verticillatum*[284] and *Hydrilla verticillata.*[283]

IV. PHYSIOLOGICAL ASPECTS OF SPATIAL ORGANIZATION OF STRUCTURES INVOLVED IN SPREAD AND VEGETATIVE REPRODUCTION

Spread, and vegetative reproduction that may result from it, is accomplished by specialized shoots and roots. The adaptive value of such organs lies in their extraordinary capacity to follow the topography of the terrain and, thus, produce their regeneration buds (axillary or adventitious) within a more or less constant range of distances above or below the soil-air interface (water-air in the case of aquatic plants). Such stems and roots are generally considered as diageotropic or plagiotropic, and we have used this terminology in previous sections of this chapter. However, it is obvious that regulation of growth of such specialized organs strictly by the angle made with the direction of the gravitational field could be self-defeating in all but strictly level terrain. Under such conditions, regeneration buds may be formed at soil depths that are too great for adventitious rooting. It is, therefore, quite likely that such organs are equipped with mechanisms that enable them (1) to sense the distance above or below the soil-air interface and (2) to change their direction of growth so as to maintain this distance within certain specific limits. The situation appears to differ in the aerial and in the subterranean environments, respectively. In the former, the most significant gradient to which growth could respond is the one in photon-flux density. Superficial stems (stolons, runners, etc.) may follow the soil contour by negative phototropism, growth into the soil being prevented by mechanical resistance. Such a mechanism has been described in ivy (*Hedera helix*), by which the stem becomes appressed against opaque objects such as walls, tree trunks, etc.[293] In the absence of light, negatively phototropic shoots would grow upwards in the direction determined by their negative gravitropic response. This is the case in a number of stoloniferous and prostrate species (*Alternanthera ficoides*, *Commelina cayennensis*, *Echinocloa colonnum*, *Portulaca oleracea*) in which the prostrate habit of the shoots was maintained only so long as it was exposed to light, while in darkness it became negatively geotropic.[294] Negative phototropism has also been invoked as the mechanism controlling the direction of growth of subterranean shoots (*Aegopodium podagraria*,[295] *Polygonatum multiflorum*, and *Urtica dioica*[296]) and roots (*Convolvulus arvensis*[297]) that are presumably much more light sensitive than their aerial counterparts. On the other hand, apices of subterranean diageotropic shoots may become negatively geotropic upon exposure to light (*Poa pratensis*,[157] *Cynodon dactylon*[158,214]). Moreover, in *C. dactylon*, when apices of erect tillers or of prostrate runners are darkened, they are transformed to the rhizomatous developmental pattern, which is quite distinctive from either of them.[214]

In contrast, the direction of growth of stolons of *C. dactylon*, *Paspalum vaginatum*, and *Stenotaphrum secundatum* was apparently determined by their geotropic response, and was unaffected by exposure to sunlight or shading. In these plants the stolons became negatively geotropic when the erect leafy shoot from which they had branched was shaded.[294] The geotropic response of these stolons is, therefore, correlatively determined by the irradiance to which their attached "parent" shoot is exposed. Correlative effects by the "parent" leafy shoot have also been reported, for instance, in *Agropyron repens*[151] and *Poa pratensis*,[157] where shading or defoliation of the "parent" shoot, as well as disconnection from it, resulted in transformation of the rhizome apex into a negatively geotropic shoot. Decapitation of the main shoot of *Ajuga reptans* results in transformation of the stolon apices into erect leafy shoots, and this is reversed by application of IAA to the cut surface.[298] In *Stachys silvatica*, subterranean stolons respond differently than epigeous stolons to detopping of their "parent" shoot. Only the former are transformed into erect shoots, while the latter retain their prostrate growth habit.[299] Spatial orientation of the rhizomes of *Cyperus rotundus* may be under hormonal control by the leafy "parent" shoot. Foliar application of GA and cytokinins (only in combination) caused the attached rhizomes to grow upwards and transform into leafy shoots with basal bulbs upon approaching the soil surface.[300]

Under the soil surface there are a number of environmental factors, in addition to light, that exhibit a pronounced gradient with increasing depth: photon-flux density, oxygen concentration, and the amplitude in the diurnal temperature fluctuations decrease with depth, while CO_2 concentration and soil moisture content increase. Moreover, mechanical resistance of the soil matrix may increase with depth and the resulting mechanical stress on the thrusting apex may release ethylene[301] with morphogenetic and probably, also, geotropic consequences. Effects of CO_2 concentration on rhizome orientation have been reported by Bennet-Clark and Ball.[295] It is quite doubtful that the gradient in these factors across the diameter of the rhizome, stolon, or root are sufficiently steep to cause differential growth rates on their upper and lower sides. It must, therefore, be assumed that if any of these factors play an active role in contour-following growth, it must be by modifying the otherwise geotropically controlled orientation. It is also likely that light may be playing different roles above and below the soil surface. In the former it may be either phototropic (gradient across the organ) or modifying the inherent geotropic response, while in the latter it most probably only acts in the modifying capacity.

Transformation of a shoot apex from a diageotropic, or rather a contour-following growth, habit to that of an erect leafy shoot is frequently a synchronous, seasonal event. Seasonality may be mediated by correlative effects of the "parent" shoot, in response to seasonal environmental signals, such as photoperiod, thermoperiod, light intensity, etc. These effects may be direct, but are more probably an indirect consequence of changes in the developmental pattern of the "parent" shoot (flowering, dormancy status) or in its growth rate (availability of assimilates, growth regulators, etc.) that are seasonal in nature. However, the possibility must be kept in mind that some environmental parameters that change seasonally, such as temperature, thermoperiod, and soil moisture stress, may directly act on the subterranean organs and transform their growth pattern. These possibilities have not received their due experimental attention.

REFERENCES

1. **Holzner, W.,** Concepts, categories and characteristics of weeds, in *Biology and Ecology of Weeds,* Holzner, W. and Numata, N., Eds., Junk Publishers, The Hague, 1982, 3.
2. **Wapshere, A. J.,** Biological control of weeds, in *Biology and Ecology of Weeds,* Holzner, W. and Numata, M., Eds., Junk Publishers, The Hague, 1982, 47.
3. **Abrahamson, W. G.,** Demography and vegetative reproduction, in *Demography and Evolution in Plant Population,* Solbrig, O., Ed., University of California Press, Berkeley, 1980, 89.
4. **Hakansson, S.,** Multiplication, growth and persistence of perennial weeds, in *Biology and Ecology of Weeds,* Holzner, W. and Numata, M., Eds., Junk Publishers, The Hague, 1982, 123.
5. **Leakey, R. R. B.,** Adaptive biology of vegetatively regenerating weeds, in *Advances in Applied Biology,* Vol. 6, Cooker, T. H., Ed., Academic Press, New York, 1981, 57.
6. **Holm, L. G., Plucknett, D. L., Pancho, J. W., and Herberger, J. P.,** *The World's Worst Weeds: Distribution and Biology,* University Press of Hawaii, Honolulu, 1977.
7. **Winkler, H.,** Uber parthenogenesis und apogamie im Pflanzenreiche, *Progr, Rei. Bot.,* 2, 293, 1908.
8. **Stearn, W. T.,** The use of the term clone, *J. R. Hortic. Soc. London,* 74, 41, 1949.
9. **Grant, V.,** *Plant Speciation,* Columbia University Press, New York, 1971, 271.
10. **Gustafsson, A.,** *Apomixis in Higher Plants,* Arssks., N. F. Ave., 2, 43, 1—370, University of Lund, Sweden, 1946-7.
11. **Stebbins, G. L.,** *Variation and Evolution in Plants,* Columbia University Press, New York, 1950.
12. **Nygen, A.,** Apomixis in the angiosperms, in *Encyclopedia of Plant Physiology,* Vol. 18, Ruhland, W., Ed., Springer-Verlag, Berlin, 1967, 551.
13. **Soane, I. D. and Watkinson, A. R.,** Clonal variation in populations of *Ranunculus repens, New Phytol.,* 82, 557, 1979.
14. **Hancock, J. F. and Bringhurst, R. S.,** Inter-population differentiation and adaptation in the perennial, diploid species of *Fragaria vesca* L., *Am. J. Bot.,* 65, 795, 1978.
15. **Van Loenhoud, P. J. and Duyts, H.,** A comparative study of the germination of some microspecies of *Taraxacum* Wigg., *Acta Bot. Neerl.,* 30, 161, 1981.
16. **Baker, H. G.,** Characteristics and modes of origin of weeds, in *The Genetics of Colonizing Species,* Baker, H. G. and Stebbins, G. C., Eds., Academic Press, New York, 1965, 147.
17. **Youngner, V. B.,** Observations of the ecology and morphology of *Pennisetum clandestinum, Øyton Buenos Aires,* 16, 77, 1961.
18. **Oinonen, E.,** The correlation between the size of Finnish bracken (*Pteridium aquilinum* (L.) Kuhn.) clones and certain periods of ancient history, *Acta Forest. Fenn.,* 83, 1, 1967.
19. **Gustafsson, A.,** Polyploidy, life form and vegetative reproduction, *Hereditas,* 34, 1, 1948.
20. **Salisbury, E.,** *Weeds and Aliens,* Collins, London, 1961.

21. **Warburg, E. F.,** Some taxonomic problems in weedy species, in *The Biology of Weeds*, Harper, J. L., Ed., Blackwell Scientific, Oxford, 1960, 11.
22. **Clausen, J., Keck, D. D., and Hiesey, W. M.,** Heredity of geographically and ecologically isolated races, *Am. Nat.*, 81, 114, 1947.
23. **Richards, A. T.,** Hybridization in *Taraxacum*, *New Phytol.*, 69, 1103, 1970.
24. **Gustafsson, A.,** The origin and properties of the European blackberry flora, *Hereditas*, 28, 249, 1942.
25. **Heslop-Harrison, Y.,** Cytological studies in the genus *Rubus* L. I. Chromosome numbers in the British *Rubus* flora, *New Phytol.*, 52, 22, 1953.
26. **Bergman, B.,** Meiosis in two different clones of the apomictic *Chondrilla juncea*, *Hereditas*, 36, 297, 1950.
27. **Brown, W. V. and Emery, W. H. P.,** Apomixis in the Gramineae: Panicoideae, *Am. J. Bot.*, 45, 253, 1958.
28. **Pritchard, T.,** Race formation in weedy species with special reference to *Euphorbia cyparissias* L. and *Hypericum perforatum* L., in *The Biology of Weeds*, Harper, J. L., Ed., Blackwell Scientific, Oxford, 1960, 61.
29. **Amor, R. L.,** Ecology and control of blackberry (*Rubus fruticosus* L. agg.). II. Reproduction, *Weed Res.*, 14, 231, 1974.
30. **Heslop-Harrison, J.,** Sexuality in angiosperms, in *Plant Physiology* (a treatise), Vol. 6C, Steward, F. C., Ed., Academic Press, New York, 1972, 134.
31. **Warmke, H. E.,** Apomixis in *Panicum maximum*, *Am. J. Bot.*, 41, 5, 1954.
32. **Bocher, T. W.,** Cytological and embryological studies in the amphi-apomictic *Arabis holboellii* complex, *Dan. Vidensk. Selsk Biol. Skr.*, 6, 1, 1951.
33. **Nygren, A.,** Form and biotype formation in *Calamagrostis purpurea*, *Hereditas*, 37, 519, 1951.
34. **Evans, L. T. and Knox, R. B.,** Environmental control of reproduction in *Themeda australis*, *Aust. J. Bot.*, 17, 375, 1969.
35. **Knox, R. B.,** Apomixis: seasonal population differences in a grass, *Science*, 157, 325, 1967.
36. **Knox, R. B. and Heslop-Harrison, J.,** Experimental control of apomixis in a grass of the *Andropogoneae*, *Bot. Not.*, 116, 127, 1963.
37. **Skalinska, M.,** Cytological analysis of some *Hieracium* species, subg. *Pilosella* from mountains of southern Poland, *Acta Biol. Cracov. Ser. Bot.*, 10, 127, 1970.
38. **Jervis, R. A. and Buell, M. F.,** *Acorus calamus* in New Jersey, *Bull. Torrey Bot. Club*, 91, 335, 1964.
39. **Wulff, H. D.,** Zur Zytologie, geographischen verbreitung und morphologie des kalmus, *Arch. Pharm. Berl.*, 287, 529, 1954.
40. **Barrett, S. C. H.,** Tristyly in *Eichhornia crassipes* (Mart.) Solms-Lanbach (water hyacinth), *Biotropica*, 9, 230, 1977.
41. **Barrett, S. C. H. and Forno, I. W.,** Style morph distribution in New World populations of *Eichhornia crassipes* (Mart) Solms-Lanbach, *Aquat. Botany*, 13, 299, 1982.
42. **Van der Pijl, L.,** *Principles of Dispersal in Higher Plants*, 3rd ed., Springer-Verlag, Berlin, 1982.
43. **Clarke, G. H.,** Important weeds of South Australia, No. 9 Soursob, *J. Agric. South. Aust.*, 38, 481, 1934.
44. **Galil, J.,** Vegetative dispersal of *Allium ampeloprasum*. I. Vegetative reproduction, *Isr. J. Bot.*, 14, 135, 1965.
45. **Jackson, D. I.,** A growth study of *Oxalis latifolia* H. B. K., *N.Z. J. Sci.*, 3, 600, 1960.
46. **Michael, P. W.,** Studies on *Oxalis pes-caprae* in Australia. I. Quantitative studies on oxalic acid during the life cycle of the pentaploid variety, *Weed Res.*, 5, 123, 1965.
47. **Robb, S. M.,** *Oxalis latifolia* Kunth., *New Phytol.*, 62, 75, 1963.
48. **Quimby, J. R., Kramer, N. W., Stephens, J. C., Lahr, K. A., and Karper, R. E.,** Grain sorghum production in Texas, *Bull. Res. Agric. Exp. Stn.*, No. 912, 1, 1958.
49. **Marchant, C. J.,** Evolution of *Spartina* (Gramineae). I. The history and morphology of the genus in Britain, *J. Linn. Soc. London. Bot.*, 60, 1, 1967.
50. **Forde, B. J.,** Translocation in grasses. I. Bermuda grass, *N.Z. J. Bot.*, 4, 479, 1966.
51. **Forde, B. J.,** Translocation in grasses. II. Perennial rye-grass and couch grass, *N.Z. J. Bot.*, 4, 496, 1966.
52. **Nyahoza, F., Marshall, C., and Sagar, G. R.,** Assimilate distribution in *Poa pratensis* L. — a qualitative study, *Weed Res.*, 14, 251, 1974.
53. **Ginzo, H. D. and Lovell, P. H.,** Aspects of the comparative physiology of *Ranunculus bulbosus* L. and *Ranunculus repens* L. II. Carbon dioxide assimilation and distribution of photosynthates, *Ann. Bot.*, 37, 765, 1973.
54. **Sarukhan, J.,** Studies on plant demography: *Ranunculus repens* L., *R. bulbosus* L. and *R. acris* L. II. Reproductive strategies and seed population dynamics, *J. Ecol.*, 62, 151, 1974.
55. **Turkington, R., Cahn, M. A., Vardy, A., and Harper, J. L.,** The growth, distribution and neighbour relationships of *Trifolium repens* in permanent pasture. III. The establishment and growth in natural perturbed sites, *J. Ecol.*, 67, 231, 1979.

56. **Schellner, R. A., Newell, S. T., and Solbrig, O. T.,** Studies on the population biology of the genus *Viola*. IV. Spatial patterns of ramets and seedlings in three stoloniferous species, *J. Ecol.*, 70, 273, 1982.
57. **Thomas, A. G. and Dale, H. M.,** The role of seed reproduction in the dynamics of established populations of *Hieracium floribundum* and a comparison with that of vegetative reproduction, *Can. J. Bot.*, 53, 3022, 1975.
58. **Tripathi, R. S.,** Ecology of *Cyperus rotundus* L. III. Populations of tubers at different depths of soil and their sprouting response to air-drying, *Proc. Natl. Acad. Sci. India (Section B)*, 39, 140, 1969.
59. **Friesen, G. H., and Hamill, A. S.,** Sprouting and development of yellow nutsedge tubers, *Can. J. Plant Sci.*, 57, 509, 1977.
60. **Thomas, P. E. L.,** Effects of desiccation and temperature on survival of *Cyperus esculentus* tuber and *Cynodon dactylon* rhizomes, *Weed Res.*, 9, 1, 1969.
61. **Grummer, G.,** Das verhalten von rhizomen der queeke (*Agropyron repens*) gegn trockene luft, *Weed Res.*, 3, 44, 1963.
62. **Forsberg, D. E.,** Another look at the Canada thistle root system, *Proc. Can. Soc. Agron.*, 8, 94, 1962.
63. **Estelita-Teixeira, M. E.,** Shoot anatomy of three bulbous species of *Oxalis*, *Ann. Bot.*, 49, 805, 1982.
64. **Selleck, G. W.,** Life history of leafy spurge, *Proc. 15th North Central Weed Control Conf.*, Weed Science Society of America, Champaign, Ill., 1958, 16.
65. **Bakker, D.,** A comparative life history study of *Cirsium arvense* (L.) Scop. and *Tussilago farfara* L., the most troublesome weeds in the newly reclaimed polders of the former Zuiderzee, in *The Biology of Weeds*, Harper, J. L., Ed., Blackwell Scientific, Oxford, 1960, 205.
66. **Zilk, S.,** Some Aspects of the Ecological Life Cycle of *Linaria vulgaris* Hill., M.Sc. thesis, University of Saskatchewan, Saskatoon, 1954.
67. **Cuthbertson, E. G.,** *Chondrilla juncea* in Australia. IV. Root morphology and regeneration from root fragments, *Aust. J. Exp. Agric.*, 12, 528, 1972.
68. **Horowitz, M.,** Development of *Cynodon dactylon* (L.) Pers., *Weed Res.*, 12, 207, 1972.
69. **Moreira, I. and Rosa, M. L.,** Estabelicimento de plantas de *Cynodon dactylon* (L.) Pers., *III. Simp. Nac. Herbologia, Portugal*, 1980, 143.
70. **Palmer, J. H.,** Studies in the behaviour of the rhizome of *Agropyron repens* (L.) Beauv. I. The seasonal development and growth of the parent plant and rhizome, *New Phytol.*, 57, 145, 1958.
71. **Yeo, R. R.,** Life history of common cattail, *Weeds*, 12, 284, 1964.
72. **Nygren, A.,** Studies on vivipary in the genus *Deschampsia*, *Hereditas*, 35, 27, 1949.
73. **Nygren, A. and Almgard, G.,** On the experimental control of vivipary in *Poa*, *Lantbrukhoegsk. Ann.*, 28, 27, 1962.
74. **Youngner, V. B.,** Environmental control of initiation of the inflorescence reproductive structures, and proliferations in *Poa bulbosa*, *Am. J. Bot.*, 47, 753, 1960.
75. **Davis, F. S. and Peters, E. J.,** Reproductive cycles of wild garlic and nomenclature of plant and bulb types, *Weeds*, 13, 84, 1965.
76. **Hakansson, S.,** *Allium veneale* L. as a weed with special reference to the conditions in south-eastern Sweden, *Vaxtodling*, 19, 1, 1963.
77. **Salisbury, E.,** *The Reproductive Capacity of Plants*, C. Bell & Sons, London, 1942.
78. **Hammerton, J. L.,** Experiments with *Cyperus rotundus* L. I. Growth and development and effects of 2,4-D and paraquat, *Weed Res.*, 14, 365, 1974.
79. **Smith, E. V. and Fick, J. L.,** Nut grass eradication studies. I. Relation of the life history of nut grass *Cyperus rotundus* L. to possible method of control, *J. Am. Soc. Agron.*, 29, 1007, 1937.
80. **Tisdale, E. W., Hironaka, M., and Pringle, W. C.,** Observations on the autoecology of *Hypericum perforatum*, *Ecology*, 40, 54, 1959.
81. **Putwain, P. D., Machin, D., and Harper, J. L.,** Studies in the dynamics of plant populations. II. Components and regulation of a natural population of *Rumex acetosella*, *J. Ecol.*, 56, 421, 1968.
82. **Hawthorne, W. R.,** The biology of Canadian weeds. IV. *Plantago major* and *P. rugellii*, *Can. J. Plant Sci.*, 54, 383, 1974.
83. **Sagar, G. R.,** Factors controlling the size of plant populations, *Proc. 10th Br. Weed Control Conf.*, 1970, 965.
84. **Ogden, J.,** The reproductive strategy of higher plants. II. The reproductive strategy of *Tussilago farfara* L., *J. Ecol.*, 62, 291, 1974.
85. **Werner, P. A.,** Competition and coexistence of similar species, in *Topics in Plant Population Biology*, Solbrig, O. T., Jain, S., Johnson, G. B., and Roxen, P., Eds., Columbia University Press, New York, 1976, 287.
86. **Abrahamson, W. G.,** Reproductive strategies in dewberries, *Ecology*, 56, 721, 1975.
87. **Holler, L. C. and Abrahamson, W. G.,** Seed and vegetative reproduction in relation to density in *Fragaria virginiana* (Rosaceae), *Am. J. Bot.*, 64, 1003, 1977.
88. **Williams, R. D.,** Intraspecific competition of yellow nutsedge, *Proc. So. Weed Sci. Soc. Am.*, 34, 231, 1981.

89. **Williams, R. D., Quimby, P. C., and Frick, K. E.,** Intraspecific competition of purple nutsedge (*Cyperus rotundus*) under greenhouse conditions, *Weed Sci.*, 25, 477, 1977.
90. **Thompson, D. H. and Beattie, A. J.,** Density-mediated seed and stolon production in *Viola* (Violaceae), *Am. J. Bot.*, 68, 383, 1981.
91. **Richards, J. H.,** Developmental potential of axillary buds of water hyacinth, *Eichhornia crassipes* Solms. (Pontederiaceae), *Am. J. Bot.*, 69, 615, 1982.
92. **Itoh, K. and Kusanagi, T.,** Growth responses of five species of Japanese arrowhead weeds to the levels of light intensity, *Proc. 8th Asian-Pacific Weed Sci. Soc. Conf.*, 1981, 307.
93. **Keely, P. E. and Thullen, R. J.,** Light requirements of yellow nutsedge (*Cyperus esculentus*), *Weed Sci.*, 26, 10, 1978.
94. **Riopel, J. L.,** Regulation of lateral root position, *Bot. Gaz.*, 130, 80, 1969.
95. **Heslop-Harrison, Y.,** Natural and induced rooting of the stem apex in *Rubus*, *Ann. Bot.*, 23, 307, 1959.
96. **Fisher, J. E.,** Evidence of circumnutational growth movements of rhizomes of *Poa pratensis* L. that aid in soil penetration, *Can. J. Bot.*, 42, 283, 1964.
97. **Fail, H.,** The effects of rotary cultivation on the rhizome weeds, *J. Agric. Eng. Res.*, 1, 68, 1956.
98. **Korsmo, E.,** *Unkrauter im Ackerban der Neuzeit*, Springer, Berlin, 1930.
99. **Bendixen, L. E.,** Anatomy and sprouting of yellow nutsedge tubers, *Weed Sci.*, 21, 501, 1973.
100. **Tumbleson, M. E. and Kommedahl, T.,** Reproductive potential of *Cyperus esculentus* by tubers, *Weeds*, 9, 646, 1961.
101. **Hauser, E. W.,** Development of purple sedge under field conditions, *Weeds*, 10, 315, 1962.
102. **Williams, R. D.,** Growth reproduction of *Cyperus esculentus* L. and *Cyperus rotundus* L., *Weed Res.*, 22, 149, 1982.
103. **Wills, G. D. and Briscoe, G. A.,** Anatomy of purple nutsedge, *Weed Sci.*, 18, 631, 1970.
104. **Wills, G. D., Hoagland, R. E., and Paul, R. M.,** Anatomy of yellow nutsedge, *Weed Sci.*, 28, 432, 1980.
105. **Wang, H. C.,** A preliminary report on the morphological studies of the vegetative body and the organization of the seed in *Nelumbo nucifera* Gaertn., *Acta Bot. Sin.*, 5, 425, 1956.
106. **Wood, C. E.,** The genera of the Nymphaceae and Ceratophyllaceae in the southwestern United States, *J. Arnold Arbor. Harv. Univ.*, 40, 94, 1959.
107. **de Wilde, W. J. J. O.,** The morphological evaluation and taxonomic value of the spathe in *Najas*, with descriptions of three new Asiatic-Malaysian taxa, *Acta Bot. Neerl.*, 10, 164, 1961.
108. **Sculthorpe, C. D.,** *The Biology of Aquatic Vascular Plants*, Edward Arnold, London, 1971.
109. **Arber, A.,** *Water Plants: a Study of Aquatic Angiosperms*, University Press, Cambridge, 1920.
110. **Yeo, R. R.,** Life history of Sago pondweed, *Weeds*, 13, 314, 1965.
111. **Borg, P. J. V.,** Ecology of *Equisetum palustre* in Finland with special reference to its role as a noxious weed, *Ann. Bot. Fenn.*, 8, 93, 1971.
112. **Mukulo, J.,** Studies on the biology and control of marsh horsetail (*Equisetum palustre* L.), *Ann. Agric. Fenn.*, 2 (Suppl. 4), 57, 1963.
113. **Williams, E. D.,** Studies on the depth distribution and on the germination and growth of *Equisetum arvense* L. (field horsetail) from tubers, *Weed Res.*, 19, 25, 1979.
114. **Raju, M. V. S., Coupland, R. T., and Stevens, T. A.,** On the ocurrence of root buds on perennial plants in Saskatchewan, *Can. J. Bot.*, 44, 33, 1966.
115. **Marsden-Jones, E. M.,** *Ranunculus ficaria:* life history and pollination, *J. Linn. Soc. London Bot.*, 50, 39, 1935.
116. **Gluck, H.,** *Biologische und morphologische Untersuchungen uber Wasser — und Sumpfgewachse*, Vol. 2, Gustav Fischer, Jena, 1906, 95.
117. **Dore, J.,** Physiology of regeneration in Cormophytes, in *Encyclopedia Plant Physiology*, Vol. 15/II, Rhuland, W., Ed., Springer-Verlag, Berlin, 1965, 1.
118. **Coupland, R. T. and Alex, J. F.,** Distribution of vegetative buds on the underground parts of leafy spurge (*Euphorbia esula* L.), *Can. J. Agric. Sci.*, 35, 76, 1955.
119. **Evetts, L. L. and Burnside, O. C.,** Root distribution and vegetative propagation of *Asclepias syriaca* L., *Weed Res.*, 14, 283, 1974.
120. **Peterson, R. L.,** The initiation and development of root buds, in *The Development and Function of Roots*, Torrey, J. G. and Clarkson, D. T., Eds., Academic Press, New York, 1975, 125.
121. **Myers, G. A., Beasley, C. A., and Derscheid, L. A.,** Anatomical studies of *Euphorbia esula* L., *Weeds*, 12, 291, 1964.
122. **Bonnett, H. T. and Torrey, T. G.,** Comparative anatomy of endogenous bud and lateral root formation in *Convolvulus arvensis* roots cultured *in vitro*, *Am. J. Bot.*, 53, 495, 1966.
123. **Hamdoun, A. M.,** The anatomy of subterranean structures of *Cirsium arvense*, *Weed Res.*, 10, 284, 1970.
124. **Schirman, R. and Zamora, B. A.,** Bud development in excised roots of rush skeleton weed (*Chondrilla juncea* L.), *Weed Sci.*, 26, 582, 1978.

125. **Bakshi, T. S. and Coupland, R. T.**, An anatomical study of the subterranean organs of *Euphorbia esula* in relation to its control, *Can. J. Bot.*, 37, 613, 1959.
126. **Charlton, W. A.**, The root system of *Linaria vulgaris* Mill. I. Morphology and anatomy, *Can. J. Bot.*, 44, 1111, 1966.
127. **Beijerinck, M. W.**, Beobachtungen und Betrachtungen Uber Wurzelknospen und Nebenwurzeln, *Natuurk. Verh. Kom. Akad. Wet. C. (Amsterdam)*, 25, 1, 1887.
128. **Priestley, J. H. and Swingle, C. F.**, Vegetative propagation from the standpoint of plant anatomy, *USDA Tech. Bull.*, 151, 1, 1929.
129. **Emery, A. E. H.**, The formation of buds on the roots of *Chamaenerion angustifolium* (L.) Scop., *Phytomorphology*, 5, 139, 1955.
130. **Dore, J.**, Studies in the regeneration of horseradish. I. A re-examination of morphology and anatomy of regeneration, *Ann. Bot.*, 19, 127, 1955.
131. **Bakshi, T. S. and Coupland, R. T.**, Vegetative propagation in *Linaria vulgaris*, *Can. J. Bot.*, 38, 243, 1960.
132. **Lund, S. and Rostrup, E.**, Marktidselem *Cirsium arvense*, en monografi, *K. Dan. Vidensk. Selsk. Skr. Naturvidensk. Math. Afd.*, 10, 1, 1901.
133. **Frazier, J. C.**, Nature and rate of development of root system of *Lepidium draba*, *Bot. Gaz.*, 105, 244, 1943.
134. **Frazier, J. C.**, Nature and rate of development of root system of *Centaurea picris*, *Bot. Gaz.*, 105, 345, 1944.
135. **Davidson, J. G.**, The establishment of *Convolvulus arvensis* in a non-competitive situation, *Proc. 10th Br. Weed Control Conf.*, 1970, 352.
136. **Frazier, J. C.**, Nature and rate of development of root system of *Convolvulus arvensis*, *Bot. Gaz.*, 104, 417, 1943.
137. **Raju, M. V. S., Stevens, T. A., and Coupland, R. T.**, Developmental studies on *Euphorbia esula* L. Morphology of the root system, *Can. J. Bot.*, 41, 579, 1963.
138. **Tschirch, A.**, Uber die Heterorhizi bei Dikotylen, *Flora*, 94, 68, 1905.
139. **Frazier, J. C.**, Nature and development of root system of *Apocynum cannabinum* L., *Bot. Gaz.*, 105, 463, 1944.
140. **Frazier, J. C.**, Nature and development of root system of *Gonolobus laevis* Michx., *Bot. Gaz.*, 106, 325, 1945.
141. **Soteres, J. and Murray, D. S.**, Root distribution and reproductive biology of honeyvine milkweed *(Cynanchum laeve)*, *Weed Sci.*, 30, 158, 1982.
142. **Pavlychenko, T. K.**, Herbicidal action of chemicals on perennial weeds, *Sci. Agric.*, 23, 410, 1943.
143. **Sagar, G. R. and Rawson, H. M.**, The biology of *Cirsium arvense* (L.) Scop., *Proc. 7th Br. Weed Control Conf.*, 1964, 553.
144. **Warwick, S. I. and Briggs, D.**, The genecology of lawn weeds. VI. The adaptive significance of variation in *Achillea millefolium* L. as investigated by transplant experiments, *New Phytol.*, 85, 451, 1980.
145. **Watt, A. S.**, Contribution to the ecology of bracken (*Pteridium aquilinum*). I. The rhizome, *New Phytol.*, 39, 401, 1940.
146. **Amor, R. L. and Harris, R. Y.**, Distribution and seed production of *Cirsium arvense* (L.) Scop. in Victoria, Australia, *Weed Res.*, 15, 407, 1975.
147. **Flamm, E.**, Zur lebensdauer und anatomie einiger rhizome, *Sitzungber. Akad. Wiss. Wien. Math. Naturwiss K. Abt.*, 131, 7, 1922.
148. **Qureshi, F. A. and Spanner, D. C.**, The effect of nitrogen on the movement of tracers down the stolon of *Saxifraga sarmentosa*, with some observations on the influence of light, *Planta*, 110, 131, 1973.
149. **Rogan, P. G. and Smith, D. L.**, Patterns of translocation of ^{14}C-labelled assimilates during vegetative growth of *Agropyron repens* (L.) Beauv., *Z. Pflanzenphysiol.*, 73 (S), 405, 1974.
150. **Fyske, H.**, Research on *Sonchus arvensis* L. I. Translocation of labelled assimilates, *Weed Res.*, 14, 305, 1974.
151. **McIntyre, G. I.**, Studies on bud development in the rhizome of *Agropyron repens*. I. The influence of temperature, light intensity and bud position on the pattern of development, *Can. J. Bot.*, 48, 1903, 1970.
152. **Qureshi, F. A and McIntyre, G. I.**, Apical dominance in the rhizome of *Agropyron repens*: the influence of nitrogen and humidity on the translocation of ^{14}C-labelled assimilates, *Can. J. Bot.*, 57, 1229, 1979.
153. **McIntyre, G. I.**, Some effects of the nitrogen supply on the growth and development of *Agropyron repens* (L.) Beauv., *Weed Res.*, 5, 1, 1965.
154. **McIntyre, G. I.**, Apical dominance in the rhizome of *Agropyron repens*. Evidence of competition for carbohydrate as a factor in the mechanism of inhibition, *Can. J. Bot.*, 47, 1189, 1969.
155. **Rogan, P. G. and Smith, D. L.**, Experimental control of bud inhibition in rhizomes of *Agropyron repens* (L.) Beauv., *Z. Pflanzenphysiol.*, 78 (S), 113, 1976.
156. **Leakey, R. R. B., Chancellor, R. J., and Vince-Prue, D.**, Parental factors in dominance of lateral buds on rhizomes of *Agropyron repens* (L.) Beauv., *Planta*, 123, 267, 1975.

157. **Nyahoza, F., Marshall, C., and Sagar, G. R.,** Some aspects of the physiology of the rhizome of *Poa pratensis* L., *Weed Res.*, 14, 329, 1974.
158. **Moreira, I.,** Tropismos dos rizomas de *Cynodon dactylon* (L.) Pers., *III Simp. Nac. Herbologia Portugal,* 1980, 198.
159. **McIntyre, G. I.,** Apical dominance in the rhizome of *Agropyron repens*. Some factors affecting the degree of dominance in isolated rhizomes, *Can. J. Bot.*, 49, 99, 1971.
160. **Sachs, T. and Thimann, K. V.,** The role of auxins and cytokinins in the release of buds dominance, *Am. J. Bot.*, 54, 136, 1967.
161. **McIntyre, G. I.,** Studies on bud development in the rhizome of *Agropyron repens*. II. The effect of nitrogen supply, *Can. J. Bot.*, 50, 393, 1972.
162. **Schwanitz, F.,** Beitrage zur analyse der pflanzen lichen polariatat, *Beih. Bot. Zbl.*, 54 (A), 520, 1936.
163. **Beasley, C. A.,** Development of axillary buds from Johnsongrass rhizomes, *Weed Sci.*, 18, 218, 1970.
164. **Hull, R. T.,** Germination control of Johnsongrass rhizome buds, *Weed Sci.*, 18, 118, 1970.
165. **Moreira, I.,** Dominancia apical nos rizomas de *Cynodon dactylon* (L.) Pers. e de *Panicum repens* L. Efeito da gravidade, temperatura e nutricao azotasa, *An. Inst. Super. Agron. Univ. Tec. Lisboa,* 39, 107, 1980.
166. **Moreira, I.,** Reorganizacao da dominancia apical em fragmentos de rhizomas de grama e escalracho, *III Simp. Nac. Herbologia, Portugal,* 1980, 185.
167. **Leakey, R. R. B., Chancellor, R. J., and Vince-Prue, D.,** Regeneration from rhizome fragments of *Agropyron repens*. II. The breaking of late spring dormancy and the influence of chilling and node position on growth from single-node fragments, *Ann. Appl. Biol.*, 87, 433, 1977.
168. **Moreira, I. and Rosa, M. L.,** Influencia da posicao do no no abrolhamento de gemas de rizoma de *Cynodon dactylon, II Simp. Nac. Herbologia, Portugal,* 1976, 23.
169. **Meyer, R. E. and Buchholtz, K. P.,** Some factors affecting the activity and growth of buds of quackgrass rhizomes, *Proc. 17th North Central Weed Control Conf.*, 1960, 37.
170. **Nigam, S. N. and McIntyre, G. I.,** Apical dominance in the rhizome of *Agropyron repens*. The relation of amino acid composition to bud activity, *Can. J. Bot.*, 55, 2001, 1977.
171. **Leakey, R. R. B., Chancellor, R. J., and Vince-Prue, D.,** Regeneration from rhizome fragments of *Agropyron repens* (L.) Beauv. I. The seasonality of shoot growth and rhizome reserves in single-node fragments, *Ann. Appl. Biol.*, 87, 423, 1977.
172. **Bourdot, G. W., Field, R. J., and White, J. G. H.,** Yarrow: numbers of buds in the soil and their activity on rhizome fragments of varying length, *N.Z. J. Exp. Agric.*, 10, 63, 1982.
173. **McWhorter, C. G.,** Factors affecting Johnsongrass rhizome production and germination, *Weed Sci.*, 20, 41, 1972.
174. **Chancellor, R. J.,** The development of dominance amongst shoots arising from fragments of *Agropyron repens* rhizomes, *Weed Res.*, 14, 29, 1974.
175. **Henson, I. E.,** Studies on the regeneration of perennial weeds in the glasshouse. II. Tropical species, *Agric. Res. Council Weed Res. Org. Oxford Tech. Rep.*, p.14, 1970.
176. **Chancellor, R. J.,** The occurrence and growth of re-inhibited shoots and dormant buds on fragmented rhizomes of *Agropyron repens* (L.) Beauv., *Proc. 9th Br. Weed Control Conf.*, 1968, 125.
177. **Chancellor, R. J.,** Further shoot regrowth from rhizome fragments of *Agropyron repens* after loss of the dominant shoot, in *Symp. on Biology and Control of Grass Weeds in Europe,* Vol. 1, Columa, 1975, 69.
178. **Koller, D., Kigel, J., Nir, I., and Ofir, M.,** Environmental Control of Weed Physiology, Final Report, U.S. Department of Agriculture Project A10-CR-27, 1975.
179. **Leakey, R. R. B. and Chancellor, R. J.,** The effects of gravity and temperature upon the reassertion of dominance in fragmented rhizomes of *Agropyron repens* (L.) Beauv., *Proc. 11th Br. Weed Control Conf.*, 1972, 50.
180. **Musik, T. Z. and Cruzado, H. J.,** The effect of 2,4-D on sprout formation in *Cyperus rotundus, Am. J. Bot.*, 40, 507, 1953.
181. **Wareing, P. F. and Nasr, T. A. A.,** Gravimorphism in trees. I. Effects of gravity on growth and apical dominance in fruit trees, *Ann. Bot.*, 25, 321, 1961.
182. **Leakey, R. R. B., Chancellor, R. J., and Vince-Prue, D.,** Regeneration from rhizome fragments of *Agropyron repens* (L.) Beauv. III. Effects of nitrogen and temperature on the development of dominance among shoots of multi-node fragments, *Ann. Bot.*, 42, 197, 1978.
183. **Leakey, R. R. B., Chancellor, R. J., and Vince-Prue, D.,** Regeneration from rhizome fragments of *Agropyron repens* (L.) Beauv. IV. Effects of light on bud dormancy and the development of dominance amongst shoots on multi-node fragments, *Ann. Bot.*, 42, 205, 1978.
184. **Amor, R. L.,** Ecology and control of blackberry (*Rubus fruticosus* L. agg.). III. Responses of *R. procerus* to mechanical removal of topgrowth and to foliage applied herbicides, *Weed Res.*, 14, 239, 1974.
185. **Marston, M. E. and Village, P. J.,** Regeneration of raspberries from root cuttings in response to physical treatment of the shoots, *Hortic. Res.*, 12, 177, 1972.
186. **Eliasson, L.,** Growth regulators in *Populus tremula*. IV. Apical dominance and suckering in young plants, *Physiol. Plant,* 25, 263, 1971.

187. **McIntyre, G. I. and Hunter, J. H.,** Some effects of the nitrogen supply on growth and development of *Cirsium arvense*, *Can. J. Bot.*, 53, 3012, 1975.
188. **Hunter, T. H. and McIntyre, G. I.,** Factors affecting translocation of 2,4-D in leafy spurge, *Weed Sci.*, 22, 167, 1974.
189. **McIntyre, G. I.,** Developmental studies on *Euphorbia esula*. Evidence of competition for water as a factor in the mechanism of root bud inhibition, *Can. J. Bot.*, 57, 2572, 1979.
190. **Raju, M. V. S., Stevens, T. A., and Coupland, R. T.,** On the regeneration of root fragments of leafy spurge (*Euphorbia esula* L.), *Weed Res.*, 4, 2, 1964.
191. **Hamdoun, A. M.,** The effect of different levels of nitrogen upon *Cirsium arvense* (L.) Scop. plants grown from seeds and from root fragments, *Weed Res.*, 10, 121, 1970.
192. **McIntyre, G. I.,** Developmental studies on *Euphorbia esula*. The influence of the nitrogen supply on the correlative inhibition of root bud activity, *Can. J. Bot.*, 50, 949, 1972.
193. **Kefford, N. P. and Caso, O. H.,** Organ regeneration of excised roots of *Chondrilla juncea* and its chemical regulation, *Aust. J. Biol. Sci.*, 25, 691, 1972.
194. **Hamdoun, A. M.,** Regeneration capacity of root fragments of *Cirsium arvense*, *Weed Res.*, 12, 128, 1972.
195. **Warmke, H. E. and Warmke, G. L.,** The role of auxin in the differentiation of root and shoot primordia from root cuttings of *Taraxacum* and *Cichorium*, *Am. J. Bot.*, 37, 272, 1950.
196. **Bonnett, H. T. and Torrey, T. G.,** Chemical control of organ formation in root segments of *Convolvulus arvensis* cultured *in vitro*, *Plant Physiol.*, 40, 1228, 1965.
197. **Richardson, R. G.,** Regeneration of blackberry (*Rubus procerus* P. J. Muller) from root segments, *Weed Res.*, 15, 335, 1975.
198. **Dore, J. and Williams, W. T.,** Studies in the regeneration of horseradish. II. Correlation phenomena, *Ann. Bot.*, 20, 231, 1956.
199. **Moore, R. J.,** The biology of Canadian weeds. XII. *Cirsium arvense* (L.) Scot., *Can. J. Bot.*, 55, 1033, 1975.
200. **Torrey, J. G.,** Endogenous bud and root formation by isolated roots of *Convolvulus* grown *in vitro*, *Plant Physiol.*, 33, 258, 1958.
201. **Bonnett, H. T., Jr.,** Phytochrome regulation of endogenous bud development in root cultures of *Convolvus arvensis*, *Planta*, 106, 325, 1972.
202. **Cuthbertson, E. G.,** Regeneration of skeleton weed from root fragments, *Aust. Weeds Res.*, 1, 10, 1962.
203. **Hakansson, S. and Wallgren, B.,** Experiments with *Sonchus arvensis* L. III. The development from reproductive roots cut into different lengths and planted at different depths, with and without competition from barley, *Swed. J. Agric. Res.*, 2, 15, 1972.
204. **Mann, H. and Cavers, P. B.,** The regenerate capacity of root cuttings of *Taraxacum officinale* under natural conditions, *Can. J. Bot.*, 57, 1783, 1979.
205. **Hakansson, S.,** Experiments with *Agropyron repens* (L.) Beauv. III. Production of aerial and underground shoots after planting rhizome pieces of different lengths of varying depths, *Lantbrukhoegsk. Ann.*, 34, 31, 1968.
206. **Hongo, A., Yagisama, Y., and Fukunaga, K.,** Growth of shoots from rhizome fragments of *Agropyron repens* (L.) Beauv. in soil, *Weed Res. (Japan)*, 25, 127, 1980.
207. **Turner, D. J.,** *Agropyron repens* (L.) Beauv. — some effects of rhizome fragmentation, rhizome burial and defoliation, *Weed Res.*, 8, 298, 1968.
208. **Vengris, J.,** The effect of rhizome length and depth of planting on the mechanical and chemical control of quackgrass, *Weeds*, 10, 71, 1962.
209. **Moreira, I.,** Influencia da profundidade de enterramento de fragmentos de rhizoma na regenercao ces de plantas de *Cynodon dactylon* (L.) Pers. e de *Panicum repens* L., *Rev. Cien. Agrarias*, 2, 111, 1978.
210. **Harradine, A. R.,** The biology of African feather grass (*Pennisetum macrourum* Trin.) in Tasmania. II. Rhizome behaviour, *Weed Res.*, 20, 171, 1980.
211. **Chancellor, R. J.,** Regeneration from depth of small rhizome fragments of *Agropyron repens* (L.) Beauv., *Aegopodium podagraria* L. and *Polygonum amphibium* L., *Proc. 8th Br. Weed Control Conf.*, 1966, 533.
212. **Lolas, P. C. and Coble, H. D.,** Johnsongrass (*Sorghum halepense*) growth characteristics as related to rhizome length, *Weed Res.*, 20, 205, 1980.
213. **Soerjani, M. and Soermawoto, O.,** Some factors affecting the germination of Alang-Alang (*Imperata cylindrica*) rhizome buds, *PANS (Pest Artic. News Summ.)*, 5, 376, 1969.
214. **Nir, I.,** Physiological and Developmental Aspects of Propagation and Spreading of Bermudagrass (*Cynodon dactylon* (L.) Pers., Doctoral dissertation, Hebrew University of Jerusalem, Israel (Hebrew, English Summary), 1978.
215. **Harper, J. L.,** Biological flora of the British Isles. *Ranunculus acris*, *Ranunculus repens* and *Ranunculus bulbosus*, *J. Ecol.*, 45, 289, 1957.
216. **Abrahamson, W. G. and Gadgil, M.,** Growth from and reproductive effort in goldenrods (*Solidago*, Compositae), *Am. Nat.*, 107, 651, 1973.

217. **Myerscough, P. J. and Whitehead, F. H.,** Comparative biology of *Tussilago farfara* L., *Chamaenerion augustifoloum* (L.) Scot., *Epilobium montanum* L. and *E. adenocaulum* Hansskm. II. Growth and ecology, *New Phytol.*, 66, 785, 1967.
218. **Hartman, H. T.,** Some effects of temperature and photoperiod on flower formation and runner production in the strawberry, *Plant Physiol.*, 22, 407, 1947.
219. **Sawamura, Y.,** Photoperiodic response of a mountain form of *Polygonum thunbergi*, *Bot. Mag. (Tokyo)*, 82, 183, 1969.
220. **McIntyre, G. I.,** Environmental control of bud and rhizome development in the seedling of *Agropyron repens* (L.) Beauv., *Can. J. Bot.*, 45, 1315, 1967.
221. **Rogan, P. G. and Smith, D. L.,** The effect of temperature and nitrogen level on the morphology of *Agropyron repens* (L.) Beauv., *Weed Res.*, 15, 93, 1975.
222. **Hakansson, S.,** Experiments with *Agropyron repens* (L.) Beauv. I. Development and growth, and the response to burial at different developmental stages, *Lantbrukhoegsk. Ann.*, 33, 823, 1967.
223. **Lovell, P. H. and Booth, A.,** Stolon initiation and development in *Solanum tuberosum* L., *New Phytol.*, 68, 1175, 1969.
224. **Booth, A.,** The role of growth substances in the development of stolones, in *The Growth of the Potato*, Ivins, J. D. and Milthorpe, F. L., Eds., Butterworths, London, 1963, 99.
225. **Oyer, E. B., Gries, G. A., and Rogers, B. J.,** The seasonal development of Johnsongrass plants, *Weeds*, 7, 13, 1959.
226. **Eussen, J. H. H.,** Studies on the growth characteristics of alang-alang *(Imperata cylindrica)*, *Biotrop. Bull.*, 18, 1, 1981.
227. **Haslam, S. M.,** The development of the annual population in *Phragmites communis*, *Ann. Bot.*, 34, 571, 1970.
228. **Holttum, R. E.,** Growth habits in the monocotyledons, *Phytomorphology*, 5, 399, 1955.
229. **Kumar, D. and Wareing, P. F.,** Factors controlling stolon development in the potato plant, *New Phytol.*, 71, 639, 1972.
230. **Wooley, D. J. and Wareing, P. F.,** The role of roots, cytokinins and apical dominance in the control of lateral shoot form in *Solanum andigena*, *Planta*, 105, 33, 1972.
231. **Thompson, P. A. and Guttridge, C. G.,** Effect of gibberellic acid on the initiation of flowers and runners in the strawberry, *Nature (London)*, 184, 72, 1959.
232. **Berger, G.,** Dormancy, Growth Inhibition and Tuberization of Nutsedge (*Cyperus rotundus* L.) as Affected by Photoperiods, Ph.D. thesis, University of California, Riverside, 1966.
233. **Yip, C. P. and Sweet, R. D.,** Tuber dormancy and response of yellow nutsedge biotypes to photoperiod and atrazine, *Proc. Northeast. Weed Sci. Soc.*, 32, 60, 1978.
234. **Garg, D. K., Bendixen, L. E., and Anderson, S. R.,** Rhizome differentiation in yellow nutsedge, *Weeds*, 15, 124, 1967.
235. **Jansen, L. L.,** Morphology and photoperiodic responses of yellow nutsedge, *Weed Sci.*, 19, 210, 1971.
236. **Suge, H. and Kusanagi, T.,** Photoperiodic control of flowering and tuberization in *Cyperus serotinus*. I. Tuberization, *Weed Res. (Japan)*, 20, 8, 1975.
237. **Yamakowa, S., Kabayashi, H., and Ueki, K.,** Effects of photoperiodic treatment on growth and propagule production by arrowhead *(Sagittaria trifolia* L.), *Weed Res. (Japan)*, 26, 118, 1981.
238. **Hammerton, J. L.,** Experiments with *Cyperus rotundus* L. III. Seasonal variation in growth, *Weed Res.*, 15, 339, 1975.
239. **Williams, R. D.,** Photoperiodic effects on the reproductive biology of purple nutsedge *(Cyperus rotundus)*, *Weed Sci.*, 26, 539, 1978.
240. **Hauser, E. W.,** Establishment of nutsedge from space-planted tubers, *Weeds*, 10, 209, 1962.
241. **Patterson, D. T.,** Shading responses of purple and yellow nutsedge (*Cyperus rotundus* and *C. esculentus*), *Weed Sci.*, 30, 25, 1982.
242. **Nyahoza, F.,** Studies of the biology of *Cyperus rotundus* L. Early Growth and vegetative reproduction strategy, *East Afr. Agric. For. J.*, 39, 120, 1973.
243. **Weber, J. A. and Nooden, C. D.,** Environmental and hormonal control of turion formation in *Myriophyllum verticillatum*, *Plant Cell Physiol.*, 17, 721, 1976.
244. **Vegis, A.,** The significance of temperature and the daily light-dark period in the formation of resting buds, *Experientia*, 9, 462, 1953.
245. **Vegis, A.,** Uber den Einfluss der temperatur und taglichen Licht-Dunkel-Period auf die bildung der Ruheknospen, zugleich ein Beitrag zur Entstehung des Ruhezustandes, *Symb. Bot. Ups.*, 14, 1, 1955.
246. **Sastroutomo, S. S.,** Environmental control of turion formation in curly pondweed *(Potamogeton crispus)*, *Physiol. Plant.*, 49, 261, 1980.
247. **Perry, T. O.,** Dormancy, turion formation and germination by different clones of *Spirodela polyrhiza*, *Plant Physiol.*, 43, 1866, 1968.

248. **Perry, T. O. and Byrne, O. R.,** Turion induction in *Spirodela polyrhiza* by abscissic acid, *Plant Physiol.*, 44, 784, 1969.
249. **Patterson, D. T., Flint, E. P., and Dickens, R.,** Effect of temperature, photoperiod, and population source on the growth of cogongrass *(Imperata cylindrica)*, *Weed Sci.*, 28, 505, 1980.
250. **Evans, M. W.,** Vegetative growth, development and reproduction in Kentucky bluegrass, *Ohio Agric. Exp. Stn. Bull.*, p. 681, 1949.
251. **Hakansson, S.,** Experiments with *Agropyron repens* (L.) Beauv. VII. Temperature and light effects on development and growth, *Lantbrukhoegsk. Ann.*, 35, 953, 1969.
252. **Patterson, D. T.,** Shading effects on growth and partitioning of plant biomass in cogongrass (*Imperata cylindrica*) from shaded and exposed habitats, *Weed Sci.*, 28, 735, 1980.
253. **Jordan-Molero, J. E. and Stoller, E. W.,** Seasonal development of yellow and purple nutsedge (*Cyperus esculentus* and *C. rotundus*) in Illinois, *Weed Sci.*, 26, 614, 1978.
254. **Wills, G. D.,** Effect of light and temperature on growth of purple nutsedge, *Weed Sci.*, 23, 93, 1975.
255. **Morgan, D. C.,** Shadelight quality effects on plant growth, in *Plants and the Daylight Spectrum*, Smith, H., Ed., Academic Press, London, 1981, 205.
256. **Smith, H.,** *Phytochrome and Photomorphogenesis*, McGraw-Hill, London, 1975.
257. **Dexter, S. T.,** Response of quack-grass to defoliation and fertilization, *Plant Physiol.*, 11, 843, 1936.
258. **Hakansson, S.,** Experiments with *Agropyron repens* (L.) Beauv. II. Production from rhizome pieces of different sizes and from seeds. Various environmental conditions compared, *Lantbrukhoegsk Ann.*, 34, 3, 1968.
259. **Watkins, J. M.,** The growth habits and chemical composition of bromegrass *Bromus inermis* Leyss. as affected by different environmental conditions, *J. Am. Soc. Agron.*, 32, 527, 1940.
260. **Harrison, C. M.,** Response of Kentucky bluegrass to variations in temperature, light and fertilizing, *Plant Physiol.*, 9, 83, 1934.
261. **McIntyre, G. I.,** Apical dominance in the rhizome of *Agropyron repens*: the influence of water stress on bud activity, *Can. J. Bot.*, 54, 2747, 1976.
262. **Ginzo, H. D. and Lovell, P. H.,** Aspects of the comparative physiology of *Ranunculus bulbosus* L. and *Ranunculus repens* L. I. Response to nitrogen, *Ann. Bot.*, 37, 753, 1973.
263. **Chancellor, R. J.,** The incidence of dormancy in two-node stem and rhizome fragments of *Polygonum amphibium* L., *Weed Res.*, 7, 323, 1967.
264. **Horowitz, M.,** Seasonal development of established Johnsongrass, *Weed Sci.*, 20, 392, 1972.
265. **Vegis, A.,** Dormancy in higher plants, *Annu. Rev. Plant Physiol.*, 15, 185, 1964.
266. **Johnson, B. G. and Buckholtz, K. P.,** The natural dormancy of vegetative buds on the rhizomes of (*Agropyron repens*) quackgrass, *Weeds*, 10, 53, 1962.
267. **Lipke, W. G., Burnside, O. C., and Haskins, F. A.,** Biochemical studies of bud dormancy in tanweed, *Plant Physiol.*, 40, 251, 1965.
268. **Davis, F. S. and McCarthy, M. K.,** Effects of several factors on the expression of dormancy in western ironweed, *Weeds*, 14, 62, 1966.
269. **Manson, W. G. and Davies, F. S.,** Dormancy in western ironweed and leafy spurge, *Weeds*, 12, 238, 1964.
270. **Dore, J.,** Seasonal variation in the regeneration of root cuttings, *Nature (London)*, 172, 1189, 1953.
271. **Coupland, R. T., Selleck, G. W., and Alex, J. F.,** The reproductive capacity of vegetative buds on the underground parts of leafy spurge (*Euphorbia esula* L.), *Can. J. Agric. Sci.*, 35, 477, 1955.
272. **Rosenthal, R. N., Schirman, R., and Robocker, W. C.,** Root development of rush skeleton weed, *Weed Sci.*, 16, 213, 1968.
273. **Otzen, D. and Koridon, A. H.,** Seasonal fluctuations of organic food reserves in underground parts of *Cirsium arvense* and *Tussilago farfara*, *Acta Bot. Neerl.*, 19, 495, 1970.
274. **Arny, A. C.,** Variations in the organic reserves in underground parts of five perennial weeds from late April to November, *Minn. Agr. Exp. Stn. Tech. Bull.*, p. 84, 1932.
275. **Hastings, R. E. and Kust, C. A.,** Reserve carbohydrate storage by yellow rocket, white cockle and hoary alyssum, *Weed Sci.*, 18, 140, 1970.
276. **Aldous, A. E.,** Relations of organic food reserves to the growth of some Kansas pasture plants, *Agron. J.*, 22, 385, 1930.
277. **Teo, C. K. H., Bendixen, L. E., and Nishimoto, R. K.,** Bud sprouting and growth of purple nutsedge altered by benzyladenine, *Weed Sci.*, 21, 19, 1973.
278. **Taylorson, R. B.,** Seasonal variation in sprouting and available carbohydrate in yellow nutsedge tubers, *Weeds*, 15, 22, 1967.
279. **Thullen, R. J. and Keeley, P. E.,** Yellow nutsedge sprouting and resprouting potential, *Weed Sci.*, 23, 333, 1975.
280. **Tumbleson, M. E. and Kommedahl, T.,** Factors affecting dormancy in tubers of *Cyperus esculentus*, *Bot. Gaz.*, 126, 186, 1962.

281. **Horowitz, M.,** Growth, tuber formation and spread of *Cyperus rotundus* L. from single tubers, *Weed Res.*, 12, 348, 1972.
282. **Basiouny, F. M., Haller, W., and Garrard, L. A.,** The influence of growth regulators on sprouting *Hydrilla* tubers turions, *J. Exp. Bot.*, 29, 663, 1978.
283. **Sastroutomo, S.,** Dormancy and germination in axillary turions of *Hydrilla verticillata, Bot. Mag. Tokyo*, 93, 265, 1980.
284. **Weber, J. A. and Nooden, C. D.,** Environmental and hormonal control of turion germination in *Myriophyllum verticillatum, Am. J. Bot.*, 63, 936, 1976.
285. **Kummerow, A.,** Beitrage zur kenutnis der Ruhepriode von Winterknospen und samen, *Beitr. Biol. Pflanz.*, 34, 293, 1958.
286. **Chawdhry, A. M. and Sagar, G. R.,** Dormancy and sprouting of bulbs in *Oxalis latifolia* H. B. K. and *O. pescaprae* L., *Weed Res.*, 14, 349, 1974.
287. **Hakansson, S.,** Experiments with *Sonchus arvensis* L., *Ann. Agric. Coll. Sweden,* 35, 989, 1969.
288. **Henson, I. E.,** Studies on the regeneration of perennial weeds in the glasshouse. I. Temperate species, *Agric. Res. Council Weed Res. Org. Oxford Tech. Rep.*, p. 12, 1969.
289. **Nyahoza, F.,** Studies on the biology of *Cyperus rotundus*. II. Sprouting potential of the vegetative axillary buds, *East Afr. Agric. For. J.*, 39, 258, 1974.
290. **Fernandez Muniz, B. and Sanchez-Tames, R.,** Seasonal patterns of growth regulators in tubers of yellow nutsedge *(Cyperus esculentus), Weed Sci.*, 30, 83, 1982.
291. **Jackson, E. K. and James, A. C.,** Response of nutsedge to plant growth regulators, *Plant Physiol.*, 49 (Suppl.), 60, 1972.
292. **Teo, C. K. H., Nishimoto, R. K., and Tong, C. S.,** Bud inhibition of *Cyperus rotundus* L. tubers by inhibitors or abscissic acid and the reversal of these effects by N-6-benzyladenine, *Weed Res.*, 14, 173, 1974.
293. **Negbi, M., Zamski, E., and Zeevi, O.,** Photo- and thigmomorphogenetic control of the attachment of the ivy (*Hedera helix* L.) to its support, *Z. Pflanzenphysiol.*, 108 (S), 9, 1982.
294. **Palmer, J. H.,** The nature of growth response to sun light shown by certain stoloniferous and prostrate tropical plants, *New Phytol.*, 55, 346, 1956.
295. **Bennett-Clark, T. A. and Ball, N. G.,** The diageotropic behaviour of rhizomes, *J. Exp. Bot.*, 29, 169, 1951.
296. **Chapham, A. R.,** Studies in the depth adjustment of subterranean plant organ. I. Raunkiaer's experiment on depth perception in *Polygonatum, New Phytol.*, 44, 105, 1945.
297. **Tepfer, D. A. and Bonnett, H. T.,** The role of phytochrome in the geotropic behaviour of roots of *Convolvulus arvensis, Planta*, 106, 311, 1972.
298. **Pfirsch, E.,** Reserches sur le Conditionnement Plagiotropique chez Quelques Plants a Stolons, Doctoral dissertation, Fac. Sci., University of Strasbourg, France, 1962.
299. **Pfirsch, E.,** Determinisme de la croissance plagiotropique chez les stolons epiges de *Stachis silvatica* L. Mise in evidence de un mechanism d'auto-correlation, *Ann. Sci. Nat. Bot. Biol. Veg.*, 6, 339, 1965.
300. **Chetram, R. S. and Bendixen, L. E.,** Gibberellic acid plus cytokinins induced basal bulbs of purple nutsedge above ground, *Weed Sci.*, 22, 55, 1974.
301. **Goeschl, J. D., Rappaport, L., and Pratt, H. K.,** Ethylene as a factor regulating the growth of pea epicotyls subjected to physical stress, *Plant Physiol.*, 41, 877, 1966.

Chapter 4

COMPARATIVE ECOPHYSIOLOGY OF WEEDS AND CROPS

David T. Patterson

TABLE OF CONTENTS

I. INTRODUCTION

Approximately 6700 species of weeds plague world agriculture.[1] Of these species, 76 have been categorized as the "world's worst weeds".[2] On the other hand, only 15 species of crops supply 90% of the world's food and occupy 75% of the world's tilled land.[3] Obviously, we know more about the physiological characteristics of the 15 major crops than about those of the many species of weeds. However, even among the crops our knowledge is far from complete.[4,5] In this chapter we will consider the branch of physiology that concerns the response of organisms to their environment. This is ecophysiology.

Understanding the comparative ecophysiology of weeds and crops is important to weed scientists for a number of reasons.[6] Knowledge of their physiological response to their environment is central to understanding how weeds and crops interact under field conditions, where they will grow, and what factors determine their relative competitiveness under different environmental conditions. Knowledge of the ecophysiology of recently introduced exotic weeds is essential in predicting their potential geographic range and agronomic impact in previously uninfested areas. Because the success of chemical weed control often depends on the presence of actively growing weeds, knowledge of a species' ecophysiology is important in timing the application of herbicides and determining stages of the life cycle that are particularly vulnerable to control. In biocontrol efforts, knowledge of the ecophysiology of the target pest, the associated crops, and the potential control agent is essential to maximize the effectiveness of control.[7]

II. METHODOLOGY FOR ECOPHYSIOLOGICAL RESEARCH

A. Environmental Control

Ecophysiological studies of weeds and crops require some means of manipulating the environment in which the plants are grown. In the field, light can be modified by artificial shading, and water availability can be controlled by "rain-out" shelters or by supplemental irrigation. Soil nutrient levels can be altered with fertilization or liming treatments, and field fumigation techniques are being developed to alter the composition of the atmospheric environment on a small scale. Some control of temperature in field studies may be obtained by growing plants at different times of year or along an elevational gradient. However, these techniques introduce variables other than temperature which complicate the experimental design.

Controlled-environment facilities such as growth chambers, air-conditioned greenhouses, and phytotrons are essential for many phases of ecophysiological research. These facilities enable the researcher to study the effects of isolated environmental factors with or without interactions of other factors. The limitations and advantages of controlled-environment research have been reviewed extensively.[8-10] The major limitation of most controlled-environment facilities is the inability to provide light of "natural" quality and quantity. The general necessity of growing plants in pots of limited soil volume also is a serious problem in long-term growth studies or in studies of drought effects in which it is desirable for the soil to dry out over an extended period of time.

The most comprehensive approach to ecophysiological research combines field and controlled-environment studies. Research problems are identified in the field, and measurements of relevant environmental variables are obtained. The importance of the different variables then can be investigated in controlled-environment chambers or greenhouses. Subsequent field studies can aid in verification of the hypotheses.

B. Experimental Techniques

Experimental techniques in ecophysiological research consist chiefly of manipulation of the test plant's environment and measurement of the plant's response to environmental factors. Techniques for evaluating photosynthetic and respiratory CO_2 exchange and plant water relations are particularly useful, because these parameters are major determinants of plant growth.

1. Photosynthesis and Transpiration

CO_2 exchange can be measured by enclosing a whole plant or a single leaf in a small chamber or cuvette and determining the CO_2 concentration in a flowing air stream before and after passage through the chamber or cuvette. The CO_2 concentration is determined with an infrared gas analyzer (IRGA). These instruments are adaptable to both field and laboratory use.[11,12] Transpirational water loss can be measured similarly, by determining the water vapor concentration in the air stream before and after passage through the chamber. Temperature, light, water vapor pressure deficit, and CO_2 concentration in the chamber can be controlled independently to provide a range of environmental conditions. Leaf stomatal resistance to CO_2 and water vapor exchange can be calculated from transpiration measurements or determined directly with diffusion porometers.[13]

2. Water Status

Plant water status can be evaluated with leaf pressure chambers or "pressure bombs"[14] or thermocouple psychrometry devices.[15] These techniques provide estimates of plant tissue water potential or the free energy status of the water present in the plant tissue. Soil water potential also can be measured with thermocouple psychrometers. Alternatively, gravimetric determinations of soil moisture content can be made with subsequent calibration to soil water potential values through the use of a pressure plate. Soil moisture also can be monitored with neutron probe instruments.

In the soil-plant-atmosphere system, water movement occurs along gradients of decreasing water potential. Therefore, measurements of plant and soil water potentials are important in assessing the plant's ability to extract water from the soil. Water stress, indicated by a decline in plant tissue water potential, affects many physiological processes.[16] Further partitioning total leaf water potential into its major components, turgor pressure potential and solute potential, provides a valuable index of plant water status as related to leaf expansion and growth.[17]

3. Growth Analysis

Another technique of considerable value in ecophysiological studies is mathematical growth analysis.[18-20] This provides a means of factoring the processes of plant growth and dry matter production into basic components. Values for leaf area and dry weights of component plant parts, collected at appropriate intervals, are used to calculate relative growth rate, relative leaf area expansion rate, net assimilation rate (rate of dry matter production per unit leaf area), leaf area duration (total amount of leaf area present over a particular time period), coefficients of partitioning of plant biomass into leaves, stems, roots, and reproductive parts, and coefficients for the distribution of total plant or leaf biomass as leaf area. Formulas for these calculations are summarized in Table 1.

Various relationships among the growth analysis parameters serve as simple models of plant growth. For example, the biomass increment or dry matter produced during a particular growth interval is approximately equal to the product of the leaf area duration (LAD) and the average net assimilation rate (NAR). The relative growth rate (R_w) or amount of dry matter produced per unit of dry matter present is approximately equal to the product of NAR and the leaf area ratio (LAR). The LAR, in turn, is the product of the leaf weight ratio (LWR = grams of leaves per gram total plant weight) and the specific leaf area (SLA = leaf area per leaf weight). These simple models provide a convenient means of determining

Table 1
MATHEMATICAL GROWTH ANALYSIS FORMULAS

Calculations requiring data from a single harvest: given that W, L, R, and S = dry weights of total plants, leaves, roots, and stems, respectively, and A = leaf area, then

Leaf weight ratio (LWR) = L/W, (g/g)
Stem weight ratio (SWR) = S/W, (g/g)
Root weight ratio (RWR) = R/W, (g/g)
Specific leaf area (SLA) = A/L, (dm^2/g)
Leaf area ratio (LAR) = A/W, (dm^2/g)

Calculations requiring data from two harvests over a time interval, ΔT: given that ΔT = length of harvest interval (days), W_1 and W_2 = total plant dry weight (g) at the beginning and end of the interval, respectively, and A_1 and A_2 = total leaf area (dm^2) at the beginning and end of the interval, respectively, then

Dry matter production (DMP) = $\Delta W = W_2 - W_1$ (g)
$\Delta A = A_2 - A_1$, (dm^2)
Net assimilation rate (NAR) or rate of dry matter production per unit leaf = $(W_2/A_2 - W_1/A_1) \times \{\alpha/(\alpha - 1)\}/\Delta T$, where $\alpha = (\ln W_2 - \ln W_1)/(\ln A_2 - \ln A_1)$, (g dm^{-2} day^{-1})
Leaf area duration (LAD) or total amount of leaf area present during harvest interval = $\Delta A/(\ln A_2 - \ln A_1) \times \Delta T$, ($dm^2$ days)
Relative growth rate (R_w) = $\ln(W_2/W_1)/\Delta T$
Relative leaf area expansion rate (R_a) = $\ln(A_2/A_1)/\Delta T$

From Patterson, D. T. and Flint, E. P., *Weed Sci.*, 31, 318, 1983. With permission.

how a selected environmental factor influences plant growth, whether through influences on the amount of leaf area (LAD or LAR) or through effects on the rate of dry matter production per unit of leaf area (NAR).

Assessment of the NAR component is fairly straightforward, since it is analogous to a whole plant net photosynthetic rate. However, the leaf area component is much more complex, involving differences in rate of production and expansion of leaves as well as anatomical differences among individual leaves. The significance of the leaf area factor in plant growth analysis was reviewed by Watson.[21]

4. *Analysis of Competition*

Experimental techniques for studying plant competition also are useful in studies of weed/crop ecophysiology. The replacement series design provides a method for assessing the relative competitiveness of species grown in mixtures.[22,23] In the simplest version of the replacement series, two species, A and B, are grown in monoculture and in mixture at the same density: e.g., AA, AB, and BB. After an appropriate growth period the plants are harvested and total dry weights (or some other growth parameters) are determined. The performance of the two species in monoculture and mixture is then compared by calculating the plant relative yield (PRY). The PRY of species A in competition with species B is the total dry weight per plant of A grown in mixture with B (AB) divided by its per plant dry weight when grown in monoculture (A from AA). Similarly, the PRY of B in competition with A is the dry weight of B from AB divided by the dry weight of B from BB. For a given comparison, PRY = 1.0 indicates that the effects of intra- and interspecific competition are similar, PRY <1.0 indicates that interspecific competition is more severe than intraspecific competition, and PRY >1.0 indicates that intraspecific competition is more severe.

The relative competitiveness of A and B can be estimated by comparing their PRY values in competition with each other. If PRY of A is greater than PRY of B, then A is a stronger competitor. Also, if, as a treatment variable (e.g., temperature, CO_2, light) changes, the PRY of A increases and the PRY of B decreases, then the change in treatment has increased the competitiveness of A with respect to B.

An additional calculation, the relative yield total, or RYT, is useful in evaluating the type of interaction that occurs when two species are grown in association.[23] The RYT is simply the sum of ([PRY of A with B]/2) and ([PRY of B with A]/2). A RYT = 1.0 occurs when A and B compete for the same resources, while RYT >1.0 indicates that A and B are exploiting the resources in different ways or somehow benefitting each other so that the total yield in mixture is greater than would be expected from the yields that occur in monoculture. This could occur if the root systems of the two plants penetrate to different depths in the soil, or if one species supplies an additional resource as in a grass/legume mixture in which the legume fixes atmospheric nitrogen into a form available to the grass. Finally, a RYT <1.0 implies mutual antagonism or allelopathy between the two species.

In studying competitive interactions of weeds and crops it is often of interest to determine whether competition above ground is more important than that below ground for water and nutrients. Several experimental techniques have been described for separating the above- and below-ground components of competition.[24,25] These techniques generally involve dividing the pots or boxes in which competing plants are grown with partitions separating either the roots or shoots or both. Allelopathy may play an important role in some weed/crop interactions. This is discussed in detail in Chapter 5.

III. GENERAL CHARACTERISTICS OF WEEDS

Because our knowledge of the ecophysiology of weeds and even of most crops is incomplete, generalizations about their comparative ecophysiology must be somewhat limited.

A. Classification

First, we may consider the botanical classification of the major crops and weeds. There are many similarities because, evolutionarily, weeds and crops may have common origins. Anderson,[26] Harlan and deWet,[27] and Harlan[3,28] have speculated that some crops such as oats may have originated as weeds, whereas some weeds have evolved from ancient crops. Harlan and deWet[27] listed 33 higher plant families containing weed and/or crop species. These families comprise less than 10% of the some 350 plant families recognized in the world's flora as a whole.[29]

The 15 "crops that feed the world"[3] include species from three monocot and five dicot families (Table 2). Two of these families, Gramineae and Leguminosae, are represented by six and three species, respectively. The remaining six families (Chenopodiaceae, Convolvulaceae, Euphorbiaceae, Musaceae, Palmae, and Solanaceae) each have only one species among the major crops.

The world's worst 18 weeds[2] include species from three monocot and four dicot families (Table 2). Again, the Gramineae are well represented, with 10 of the 18 species. Three families (Gramineae, Chenopodiaceae, and Convolvulaceae) have species among both the 15 major crops and the 18 worst weeds. If we expand our weed list to include the 76 worst weeds of the world,[2] 30 plant families are represented. Three of these additional families (Euphorbiaceae, Leguminosae, and Solanaceae) also appear in the list of the world's major crops. Only eight of the 30 families have more than one species among the 76 worst weeds.

In addition to their taxonomic classification in relatively few plant families, the major agronomic weeds share a number of characteristics that are recognized as contributing to their negative impact on agricultural production. For example, Baker[30,31] proposed a list of "ideal weed characteristics" that apply to botanical weeds or "ruderals" as well as to agronomic weeds or "agrestals". Other characteristics have been summarized by Hill[32] and Patterson.[6]

As pointed out by Hill,[32] some characteristics that are of adaptive value in ruderals may be disadvantageous in agronomic weeds. For example, Baker's[30] original list of ideal weed characteristics includes "no special environmental requirements for germination." This

Table 2
PLANT FAMILIES OF AGRONOMIC IMPORTANCE

Family	Species among 15 major crops[a]	Species among 18 world's worst weeds[b]	Species among 76 world's worst weeds[b]
Amaranthaceae	0	2	2
Araceae	0	0	1
Asteraceae	0	0	10
Boraginaceae	0	0	1
Brassicaceae	0	0	1
Caryophyllaceae	0	0	2
Ceratophyllaceae	0	0	1
Chenopodiaceae	1	1	1
Commelinaceae	0	0	1
Convolvulaceae	1	1	1
Cyperaceae	0	2	5
Equisetaceae	0	0	1
Euphorbiaceae	1	0	1
Gramineae	6	10	29
Leguminosae	3	0	2
Malvaceae	0	0	1
Musaceae	1	0	0
Oxalidaceae	0	0	1
Palmae	1	0	0
Papaveraceae	0	0	1
Plantaginaceae	0	0	1
Polygonaceae	0	0	2
Pontederiaceae	0	1	2
Portulacaceae	0	1	1
Primulaceae	0	0	1
Rubiaceae	0	0	1
Salvineaceae	0	0	1
Scrophulariaceae	0	0	1
Solanaceae	1	0	1
Sphenocleaceae	0	0	1
Verbenaceae	0	0	1
Zygophyllaceae	0	0	1

[a] From Harlan.[3]
[b] From Holm et al.[2]

characteristic might prove detrimental in agronomic weeds because the successful establishment of such weeds often is keyed to factors involved in establishing the crop (see Chapter 2). Thus, a weed with restricted germination requirements might escape chemical or cultural control measures.

B. Characteristics Contributing to Success of Weeds

Characteristics that are of particular adaptive value in agronomic weeds can be grouped into three categories: (1) those related to the physiology, growth, and competitiveness of the weed; (2) those related to the reproductive biology of the weed; and (3) those related to cultural practices involved in production of the crops that the weed typically infests. These characteristics are summarized in Table 3.

1. Competitiveness

We often think of agronomic weeds as very competitive plants. Certainly, competition for light, water, and nutrients is the major factor accounting for the adverse effect of weeds on crop growth and yield. However, many of the major agronomic weeds thrive only in the

Table 3
ADAPTIVE CHARACTERISTICS OF AGRONOMIC WEEDS[a]

Related to physiology, growth, and competitiveness
- High relative growth rates in seedling stage[34]
- High rates of photosynthesis.[52] Rapid development of exploitative root systems
- Rapid partitioning of photosynthate into new leaf area production[99]
- Rapid vegetative growth to reproductive phase
- Special "weapons" for interference
- Freedom from environmental constraints — "general purpose genotype", high capacity for acclimation to changing environment

Related to reproductive phase
- Breeding systems that provide some outcrossing but also allow self-fertilization
- Copious seed production under favorable conditions with some seed production occurring over a range of favorable and stressful conditions
- Pollination by wind or generalized insect visitors

Related to cultural practices
- Morphological and physiological similarity to crop
- Timing of seed maturity to coincide with crop harvest
- Resistance or tolerance to chemical herbicides
- Resistance to mechanical control — regeneration from rhizomes or other vegetative propagules
- Seed dormancy, longevity in soil; discontinuous germination over long periods of time

[a] Adapted from Baker,[30,31] Hill,[32] and Patterson.[6] Superscript numbers refer to references.

intensively managed and regularly disturbed habitat of the cultivated field. In this environment, weeds benefit from environmental stress-reducing practices such as irrigation, fertilization, and pest control that are intended to benefit the crops. These weeds would not be considered strong competitors in the botanical sense, because they generally lack the ability to tolerate shade and to invade and survive in established vegetation. In North Carolina, for example, few major agronomic weeds are present even in the earliest successional stages following the abandonment of crop lands.[33] These annual weeds are replaced by short-lived perennial plants which, while more competitive, are less well adapted to the intense disturbance typical of cultivated fields.

In the context of Grime,[34] most agronomic weeds exhibit the "ruderal" or "competitive-ruderal" growth habit characteristic of plants from habitats with relatively low levels of environmental stress and with the severity of competition limited by frequent disturbance. Most crop plants, likewise, fit in this category. Thus, from an ecological standpoint as well as from the standpoint of evolutionary history, domesticated crops and many of their associated weeds are very similar. Highly evolved crop mimics or "satellite" weeds may be unable to survive in the absence of man's agricultural activities.[35]

Competitiveness in both crops and agronomic weeds is related to their ability to exploit and sequester the environmental resources upon which plant growth depends. Obviously, competition is important only during the crop growing season. Therefore, relative crop/weed competitiveness generally depends on plant performance during only a part of each year. Furthermore, many studies have shown that the critical period for weed/crop competition occurs during only the early part of the growing season.[36] Thus, weed/crop interactions in the agroecosystem are of shorter duration and less complex than the interactions that occur in less intensively disturbed ecosystems in which plants compete over longer periods of time.

Competitiveness is not an intrinsic property of any particular weed or crop; it is only measurable in comparison with another species.[37] However, certain characteristics tend to be associated with competitiveness, as pointed out by Clements et al.[38] quoted by Donald:[24]

It is evident that practically all the advantages or weapons of competing species are epitomized in two words — amount and rate. Greater storage in seed or rootstock, more rapid and complete germination, earlier start, more rapid growth of roots and shoots, taller and more branching stems, deeper and more spreading roots, more tillers, larger leaves, and more numerous flowers are all the essence of success.

This description of competitiveness is particularly relevant to weeds and crops because it emphasizes the resource exploitation aspect of plant competition. Larger, more vigorously growing plants with rapidly expanding leaf canopies and root systems are more efficient at utilizing the resources that are essential to plant growth. In the agroecosystem these aspects of competitiveness become more important than other characteristics such as shade tolerance and the ability to invade established vegetation.

We may consider the opportunistic agronomic weed as a "resource sponge" that occupies the ecological vacuum which is created in a crop as part of the cultural practice. In its seedling stage early in the growing season, such a plant may escape the intense competition typical of established vegetation. As the growing season progresses and the crop/weed canopy closes, competition with other weeds of the same or different species, and with the crop, becomes a limiting factor. However, the successfully adapted agronomic weed has been ushered through the critical establishment phase largely in the absence of competition because it is keyed to the cultural sequence of the crop.

2. Similarities to Crop Plants

Thus, the most critical characteristics contributing to success in many major agronomic weeds may have to do with their adaptations to many of the chemical and cultural practices associated with crop production. These adaptations may be of much greater significance than many of the characteristics attributed to botanical weeds. According to Bunting,[39] the adaptation of the weed to the crop is of paramount importance: "The more nearly a weed species resembles a crop in habit, ecological requirements, and in time relations [phenology], the more difficult it must be to control without damaging the crop itself." The phenomenon of "crop mimicry" in highly evolved weeds has been reviewed recently by Barrett.[35]

Adaptation of a weed to a particular crop may occur at all stages of the life cycle from seed or other propagule to mature plant. Similarity of weed propagules with crop seed has long been known to be of great adaptive value. When such similarity is coupled with timing of weed seed maturation to coincide with crop harvest, the chances for widespread dissemination of the weed species as a crop contaminant are greatly increased. Examples of this phenomenon include *Camelina sativa* (L.) Crantz in flax (*Linum usitatissimum* L.), *Bromus secalinus* L. in rye (*Secale cereale* L.), and *Lolium temulentum* L. in wheat (*Triticum aestivum* L.). In the southern U.S., *Cardiospermum halicacabum* L. produces a seed almost identical in size and shape to soybean (*Glycine max* L. Merr.).[40] Thus, *Cardiospermum* seed is readily harvested with soybean, and it is very difficult to mechanically separate the weed seed from the crop.

Morphological and physiological similarities between crops and their associated weeds at later stages in the life cycle also contribute to the negative impact of some weed species. An early example of this was described by Harlan.[41] At a cereal nursery, he observed that wide variation in the morphology and growth habit of *Avena fatua* L. was correlated with the growth form of the infested crop. Thus, at the same time of year, *A. fatua* L. exhibited a sprawling rosette growth form in a winter barley (*Hordeum vulgare* L.) crop, while in an adjacent crop of durum wheat (*T. durum* L.) the *A. fatua* L. plants were tall and erect. In each case the height of the *A. fatua* L. was similar to that of the crop. The time of seed maturation in the *A. fatua* L. also appeared to be correlated with the similar phenological stage of the infested crop.

The extremes of crop mimicry by weeds develop when hybridization occurs between closely related weeds and crops. For example, in sorghum, hybridization of the cultivated

Sorghum bicolor (L.) Moench with weedy *S. halepense* (L.) Pers. results in an aggressive perennial weed that produces few seed but has vigorous vegetative growth.[42] Chemical control of these weeds is extremely difficult because of their physiological similarity to the crop. Hybridization between the horticultural crop radish (*Raphanus sativus* L.) and the weed *R. raphanistrum* L. also has been reported.[43] This hybridization has resulted in a weedy form of *R. sativus* L. that has dormant seeds and a more branched and penetrating root system than the cultivated form. A similar situation is found in sugar beets (*Beta vulgaris* L.) where hybridization with the wild beet (*B. vulgaris* L. ssp. *maritima* [L.] Arc./Thell) has resulted in an annual form that fails to produce the large sugar-storing root of the cultivated beet.[44] Like the weedy sorghums and radishes, this weedy beet is extremely difficult to control chemically.

3. Morphological and Physiological Plasticity

In addition to the aspects of competitiveness and similarity to the infested crop that we have discussed, other weed characteristics in Table 3 are important in our consideration of the comparative ecophysiology of weeds and crops. One of these, "freedom from environmental constraints", is an area that has received consideration in ecophysiological studies of many plant species.

The capability for phenotypic adjustment to environmental perturbations clearly is advantageous in a plant that may be exposed to the range of environmental conditions characteristic of the cultivated field. This capability is a major aspect of the "general purpose genotype" considered by Baker[31] to be of such importance in successful weeds. Weeds with the general-purpose genotype do not exhibit the finely tuned, genetically based ecotypic differentiation typical of many nonweedy species. In contrast, they have a high degree of phenotypic plasticity that can provide a wide range of morphological characteristics and physiological responses within a limited number of genotypes. This plasticity confers "preadaptation" to a potentially wide range of habitats and growing conditions and, thus, helps free a species from environmental constraints. It should be obvious that this characteristic distinguishes weeds from most crops, in which, through intensive selection in crop breeding, much of the flexibility in response to environmental variation may have been lost.

The general-purpose genotype characteristic may be manifested at many stages during the life cycle of a successful weed. For example, dimorphism in seed is well known in *Xanthium*, in which upper and lower seeds from the same fruit have different germination requirements.[45] This adaptation helps assure the survival of seed population of *Xanthium* in agricultural soils. In other weeds, seeds from different parts of the same parent plant may differ in dormancy or in size with consequent implications for differences in survival or dispersal. Seed polymorphism is discussed in more detail in Chapter 1. Phenotypic plasticity in plant age and size at which mature, viable seeds are produced is characteristic of many weeds.[46]

Plasticity in physiological responses to environmental factors can occur through the phenomenon of acclimation, a reversible metabolic or physiological adjustment to prevailing environmental conditions. This process has been studied intensively in plants from environmentally harsh habitats such as deserts, arctic and alpine areas,[47] and, to a lesser extent, in crops and agronomic weeds.[48] Acclimation of the photosynthetic process to changes in ambient temperature and light are well known in a number of species.[48] Acclimation to water stress also occurs through osmotic adjustment.[49] In addition to these short-term reversible physiological adjustments, morphological changes in leaves and other organs also enable crops and weeds to adapt to changing environmental conditions. This morphological "acclimation" may or may not be reversible. Physiological and morphological acclimation will be discussed in greater detail in the final section of this chapter.

4. Photosynthetic Pathway

A final group of ecophysiological and morphological characteristics apparently of considerable adaptive value in agronomic weeds is the syndrome associated with the C_4 photosynthetic pathway. Shortly after the discovery and elucidation of the C_4 pathway,[50,51] Black et al.[52] pointed out that many of the major weeds of the world have this pathway, rather than the more prevalent C_3 pathway or Calvin cycle.[53] "Photosynthetically efficient" plants with the C_4 pathway were said to be more competitive than the "nonefficient" plants that possess the C_3 pathway. Since the publication of Black et al.'s[52] stimulating paper, many more species of plants have been surveyed to determine whether they have the C_3 or C_4 pathway.[54,55] Thus far, some 500 species of angiosperms from 117 genera in 13 families have been found to possess the C_4 pathway.

According to Stebbins,[29] the angiosperm flora of the world consists of 231,413 species from 12,334 genera in 349 families. Thus, only about 0.2% of the total species are known to be C_4 plants. Of course, only a relatively few species have been examined for the presence of the C_4 pathway. At higher levels of taxonomic classification, 0.95% of the angiosperm genera have at least one C_4 species, while 3.7% of the 349 families have genera containing C_4 species. A number of genera contains both C_3 and C_4 species.

Even though surveys for C_4 photosynthesis are by no means complete, existing data indicate that the pathway occurs in relatively few plants, considering the world's flora as a whole. However, as first shown by Black et al.,[52] C_4 species are disproportionately overrepresented in lists of major weeds. For example, of the 76 worst weeds in the world,[2] 42% are C_4 plants.[56] An even greater 78% of 18 worst weeds[2] of the world are C_4 plants.[56] In comparison, only 20% of the 15 crops that feed the world[3] are C_4 plants; the remaining 80% are C_3 plants. Major C_3 and C_4 crops and weeds are summarized in Table 4.

In spite of the presumed advantages of the C_4 pathway, many of the most troublesome and costly agronomic weeds lack it. Cocklebur (*Xanthium* spp. L.), morningglory (*Ipomoea* spp. L.), wild oats (*A. fatua* L.), lambsquarters (*Chenopodium album* L.), velvetleaf (*Abutilon theophrasti* Medic.), and sicklepod (*Cassia obtusifolia* L.) are examples of major C_3 weeds. Furthermore, many C_4 weeds are not competitive in the botanical sense.[57] However, as we discussed earlier, the success and impact of agronomic weeds are more closely related to their abilities to thrive in disturbed habitats in which competition is less intense than in established vegetation. The C_4 pathway and its associated physiological characteristics appear to be of adaptive value in such environments.

A detailed discussion of the C_3 and C_4 photosynthetic pathways appears in Chapter 1 of Volume II. From a functional standpoint, the C_4 photosynthetic pathway operates as a mechanism for concentrating atmospheric CO_2 at the ultimate site of fixation within the leaf.[58] In plants with the C_4 pathway, CO_2 initially is fixed into the four-carbon compound oxaloacetic acid through the carboxylation of phosphoenol-pyruvate (PEP) by PEP carboxylase (PEP_c). This fixation occurs in the cytoplasm of the mesophyll cells surrounding the vascular bundle sheath cells of the leaf. The fixed CO_2 is then transferred in the form of the four-carbon compounds aspartic or malic acid (formed from oxaloacetic acid) to chloroplasts in the bundle sheath cells where the CO_2 is released through decarboxylation of malic or aspartic acid. The CO_2 (now in the form of bicarbonate ion) then enters the C_3 photosynthetic pathway through the carboxylation of ribulose bisphosphate by ribulose bisphosphate carboxylase ($RuBP_c$). The initial stable product of this fixation is the three-carbon compound phosphoglyceric acid, hence, the "C_3 photosynthetic pathway".

The CO_2-concentrating effect of the C_4 pathway occurs because PEP_c has a much higher affinity for CO_2 than does $RUBP_c$, the enzyme responsible for the initial fixation of atmospheric CO_2 in C_3 plants. Because of this inherent CO_2-concentrating mechanism in C_4 plants, increases in the external CO_2 concentration stimulate net photosynthesis less in C_4 than in C_3 plants.

The absence of photorespiration also contributes to the higher net photosynthetic rates of C_4 plants. Photorespiration occurs in C_3 plants because $RuBP_c$ also can function as an

Table 4
PHOTOSYNTHETIC PATHWAYS OF MAJOR CROPS AND WEEDS

Crops		Weeds	
C_3	C_4	C_3	C_4
Arachis hypogaea L.	*Pennisetum glaucum*	*Abutilon theophrasti* Medic.	*Amaranthus* spp.
Avena sativa L.	*Saccharum officinarum* L.	*Ageratum* spp.	*Brachiaria* spp.
Beta vulgaris L.	*Setaria italica* (L.) Beauv.	*Agropyron repens* (L.) Beauv.	*Cenchrus* spp.
Cocos nucifera L.	*Sorghum bicolor* (L.) Moench	*Ambrosia* spp.	*Cynodon dactylon* (L.) Pers.
Glycine max (L.) Merr.	*Zea mays* L.	*Asclepias syriaca* L.	*Cyperus rotundus* L.
Gossypium hirsutum L.		*Avena fatua* L.	*C. esculentus* L.
Hordeum vulgare L.		*Bidens pilosa* L.	*Digitaria* spp.
Ipomoea batatas (L.) Lam.		*Brassica* spp.	*Echinochloa* spp.
Manihot esculenta Crantz		*Cassia obtusifolia* L.	*Eleusine indica* (L.) Gaertn.
Medicago sativa L.		*Chenopodium album* (L.)	*Euphorbia hirta* L.
Musa spp.		*Cirsium arvense* (L.) Scop.	*Imperata cylindrica* (L.) Beauv.
Oryza sativa L.		*Convolvulus arvensis* L.	*Kochia scoparia* L. Schrad.
Phaseolus vulgaris L.		*Datura stramonium* L.	*Leptochloa* spp.
Secale cereale L.		*Eichhornia crassipes* (Mart.) Solms	*Panicum dichotomiflorum* Michx.
Solanum tuberosum L.		*Galium aparine* L.	*P. maximum* Jacq.
Triticum spp.		*Ipomoea* spp.	*P. repens* L.
Zea mays L.		*Lantana camara* L.	*P. texanum* Buckl.
		Lolium temulentum L.	*Paspalum dilatatum* Poir.
		Mimosa spp.	*Pennisetum* spp.
		Plantago spp.	*Portulaca oleracea* L.
		Polygonum spp.	*Rottboellia exaltata* L.f.
		Rumex spp.	*Salsola iberica* Sennen & Pau
		Sida spp.	*Setaria* spp.
		Solanum spp.	*Sorghum halepense* (L.) Pers.
		Stellaria media (L.) Cyrill.	*Sporobulus* spp.
		Striga spp.	*Tribulus terrestris* L.
		Xanthium spp.	

oxygenase to produce phosphoglycerate, a substrate for photorespiration.[59] The oxygenase function of $RuBP_c$ is reduced at higher CO_2 levels. Therefore, the CO_2 concentrating effect of the C_4 pathway reduces or eliminates the source of phosphoglycerate. Furthermore, the radial arrangement of the leaf mesophyll cells (containing PEP_c) around the bundle sheath cells (where the C_3 pathway and photorespiration occur) results in the refixation or trapping of any CO_2 that might be released through any photorespiration that might occur.

Despite these apparent advantages, the C_4 pathway does not operate without an energy cost to the plant. This energy cost is in the form of two additional ATP molecules for every molecule of CO_2 fixed through the C_4 pathway.[60] This additional energy cost of the C_4 pathway results in a lower quantum yield in C_4 plants, in which the C_3 pathway also operates at a cost of three ATP for every molecule of CO_2 fixed. Of course, this additional energy cost in C_4 plants may be balanced by the conservation of CO_2 through reduced photorespiration.[61] In fact, Ehleringer and Bjorkman[62] have demonstrated that the relative energy costs of the C_3 and C_4 pathways vary with temperature so that at 30°C the quantum yields of C_3 and C_4 plants are essentially the same. At temperatures below 30°C, the quantum efficiency of C_3 plants exceeds that of C_4 plants while the reverse is true at higher temperatures. This difference in quantum yield has been used as a basis for explaining the effects of temperature on seasonal and geographical distributional patterns of C_3 and C_4 grasses.[63]

Other differences between C_3 and C_4 plants that relate to their ecophysiology have been reviewed previously.[64,65] In comparison with C_3 plants, C_4 plants have higher temperature optima and higher light-saturation ranges for photosynthesis. The CO_2-concentrating function of the C_4 pathway results in greater water-use efficiency on a whole-plant basis in C_4 plants, because of their higher photosynthesis/transpiration ratio. This occurs because at a given level of stomatal resistance, the internal CO_2 concentration at the site of CO_2 fixation in the leaf and, hence, the rate of CO_2 fixation are greater in C_4 than in C_3 plants. For given leaf temperatures and leaf-to-air vapor pressure differences, the transpiration rate is controlled by stomatal resistance. The net photosynthetic rate, however, depends on internal leaf factors as well. These internal factors are less limiting in C_4 plants than in C_3 plants.

The efficiency of nitrogen use also may be greater in C_4 than in C_3 plants.[66,67] This theory is based on the observation that $RUBP_c$ or fraction I protein levels are two- to fourfold lower in the leaves of C_4 grasses in comparison with C_3 grasses. The $RUBP_c$ protein is the major N-containing substance in the leaves of most C_3 plants, constituting up to 50% or more of the total soluble protein. In C_4 plants, the $RUBP_c$, in addition to being present in lower amounts, also operates in a CO_2-enriched atmosphere because of the CO_2-concentrating effect of the C_4 pathway. The primary carboxylation enzyme in C_4 plants (PEP_c) is present in low concentrations relative to $RUBP_c$. Therefore, the higher net photosynthetic rates in the leaves of C_4 plants are achieved with less investment in nitrogen-containing compounds, and the rate of CO_2 fixation per unit of leaf or total plant nitrogen content, consequently, is considerably greater. The differences between C_3 and C_4 plants in compartmentalization of nitrogen-assimilating enzymes also probably affect nitrogen use efficiency (see Chapter 1, Volume II). In field studies, C_4 grasses characteristically exhibit greater yield increases than C_3 grasses in response to nitrogen fertilization. This would indicate a higher nitrogen-use efficiency on a whole-plant or stand level corresponding to the higher rate of photosynthesis per unit of leaf nitrogen.

Based on their higher net photosynthetic rates, C_4 plants might be expected to have higher growth rates than C_3 plants. This was one basis for the speculation by Black et al.[52] that C_4 plants are more competitive than C_3 plants. As Baskin and Baskin[57] have pointed out, however, many factors other than growth rate contribute to competitive ability in plants. Furthermore, according to the analyses of Gifford[68] and Evans,[4] the superiority of C_4 over C_3 plants decreases as the level of organization or complexity of the system increases. For example, at the initial carboxylation step, PEP_c has a 60- to 70-fold greater affinity for CO_2

than does $RUBP_c$. The leaf mesophyll conductance to CO_2 (a measure of the photosynthetic potential of the intact leaf exclusive of limitation imposed by the stomata) of C_4 plants is about five times that of C_3 plants, and the single-leaf photosynthetic rate per unit area of C_4 plants is 1.5 to 3 times that of C_3 plants. Gifford[68] concluded that, under optimum environmental conditions for each group of plants, C_3 and C_4 plants may have about the same potential short-term productivity on a unit land area basis. The analyses of Gifford[68] and Evans[4] have been questioned by Monteith,[69] who showed that C_4 plants do have higher rates of production than C_3 plants, but the differences are not as great as might be predicted from their single-leaf photosynthetic rates.

In theory, the C_4 photosynthetic pathway is most likely to be of adaptive value in environments characterized by high irradiance, high temperature, and low water availability.[64,70] Studies of the climatic distribution and seasonal growth patterns of some C_3 and C_4 species seem to support this generalization. For example, in the North American short grass prairie, C_3 grasses often complete most of their growth cycle early in the season, whereas C_4 grasses are active later, during hotter and drier periods.[71] The proportion of C_4 species in the general flora normally is greater at lower latitudes[72] and lower elevations.[73] Along aridity gradients, the proportion of C_4 species tends to increase with increasing aridity.[74]

The proportion of C_4 species among the major weed species associated with particular crops similarly appears to be somewhat related to climate, particularly from the standpoint of the time of year during which the crop is grown.

In Table 5 is shown the proportion of C_3 and C_4 weeds in major crops and "agricultural use categories" in four regions of the U.S.[75] From this table it is clear that the proportion of C_4 weeds in various crops generally varies much more with the crop than with the geographical region. For example, the proportion of C_4 weed species in the warm-season crop, corn (*Zea mays* L.) (averaged over the four regions), is about 64%, whereas in wheat, a cool-season crop, only 10% of the major weeds are C_4 plants. Among other important agronomic crops, cotton (*Gossypium hirsutum* L.), peanuts (*Arachis hypogaea* L.), sugarcane (*Saccharum officinarum* L.), tobacco (*Nicotiana tabacum* L.), sorghum (*Sorghum bicolor* (L.) Moench), and sugarbeets (*B. vulgaris* L.) all have weed floras comprised of more than 50% C_4 plants, whereas about 44% of the weeds in soybeans are C_4 plants.

The proportion of C_4 weeds reported in rice (*Oryza sativa* L.) from the southern region is rather low, despite the fact that rice is a warm-season, warm-region crop. An explanation for this lower proportion of C_4 weeds in rice may be a lack of flooding tolerance in C_4 plants. Takeda et al.[76] reported a greater proportion of C_4 weeds in upland rain-fed rice than in flooded paddy rice. Grasses with the C_3 pathway predominated in flooded lowland rice. However, C_4 species in three genera (*Echinochloa*, *Leptochloa*, and *Paspalum*) tolerated flooded conditions and were important weeds even in lowland rice.

When the proportions of C_3 and C_4 weeds were calculated on a regional basis by averaging the values for the 11 major agronomic crops, little correlation with geographical region was apparent (Table 5). The C_3/C_4 proportions were 54/46 in the northeastern region, 53/47 in the north central region, 51/49 in the southern region, and 48/52 in the western region. In crops such as hay and pastures that are less frequently cultivated and, perhaps, less intensively managed for weed control, the proportion of C_4 weeds was low. This may reflect the inability of C_4 weeds to compete with more established vegetation as suggested by Baskin and Baskin.[57]

IV. EFFECTS OF ENVIRONMENT ON WEED AND CROP PHYSIOLOGY, GROWTH, AND COMPETITIVENESS

With the preceding general background in mind, we can now consider the principal effects of variations in light, temperature, water, mineral nutrients, and atmospheric CO_2 concen-

Table 5
PERCENT OF C_3 AND C_4 WEEDS REPORTED BY CROP AND REGION IN THE U.S.[a]

Crop	Northeastern		North Central		Southern		Western		Mean % C_4[b]
	C_3	C_4	C_3	C_4	C_3	C_4	C_3	C_4	
Corn	36	64	41	59	32	68	34	66	64
Cotton	—	—	40	60	41	59	47	53	57
Peanuts	—	—	—	—	45	55	20	80	68
Rice	—	—	—	—	60	40	40	60	50
Sorghum	47	53	36	64	44	56	37	63	59
Soybean	53	47	62	38	54	46	—	—	44
Sugarbeet	41	59	45	55	—	—	46	54	56
Sugarcane	—	—	—	—	0	100	40	60	80
Tobacco	40	60	33	67	39	61	—	—	63
Wheat	96	4	82	18	98	2	85	15	10
Other small grain	68	32	81	19	98	2	81	19	18
Mean % by region[c]	54	46	53	47	51	49	48	52	
Hay	95	5	79	21	66	34	84	16	19
Annual pasture	60	40	69	31	80	20	38	62	38
Perennial pasture	92	8	79	21	81	19	86	14	16

[a] Numbers in table are % of total reports.
[b] Averaged across regions.
[c] Averaged across crops.

From USDA, Extent and Cost of Weed Control with Herbicides and an Evaluation of Important Weeds, 1968, A.R.S.-H-1, November 1972; available from the NTIS, Springfield, Va.

tration on the ecophysiological responses of weeds and crops. Light, water, and mineral nutrients are the major environmental resources for which weeds and crops compete.[24,38] The availability of these resources influences the physiological, morphological, and competitive responses of all plants. Carbon dioxide also is a principal resource for plant growth; 40% of plant dry matter is carbon. Competition for CO_2 may occur in the agroecosystem if diffusion of CO_2 is restricted by the weed or crop canopy and if photosynthesis proceeds at such a rate that the CO_2 concentration falls below normal ambient levels. Temperature, while not a resource or object of competition as such, is a major factor controlling plant responses.

A. Light

Light supplies the energy for photosynthetic CO_2 fixation and controls the development and morphology of the plant through effects of quality (spectral distribution) and duration (photoperiod) as well as total energy.[77] In the presence of adequate water and nutrients and favorable temperatures, available light energy sets the limits for plant productivity.[24] Understanding the ecophysiological aspects of the effects of light on weeds and crops is important because weeds compete with crops for available light energy. In fact, promoting the development of a crop canopy is a major means of cultural weed control.

1. Spectral Quality Effects

The presence of a weed or crop canopy alters the quality or wavelength distribution as well as the quantity of the light energy passing through it. This occurs because green leaves absorb light energy in the blue and red regions of the spectrum while reflecting or transmitting energy in the green and far-red regions. Thus, in comparison with unaltered sunlight, light under a plant canopy has a much lower ratio of red light to far-red light (R:FR). The significance of this effect has been reviewed recently by Smith.[77] Many of the morphological and physiological effects of canopy shading may be associated with alterations in the R:FR which, in turn, affect the phytochrome system. For example, rapid stem elongation under canopy shade may be a response to altered R:FR, while shade-induced changes in the distribution of plant biomass as leaf area appear to be related to decreased total light energy.[77] The effects on weed seed germination associated with changes in R:FR under canopy shade are discussed in Chapter 2.

2. *Light Energy Effects*

Most weeds and crops reach their maximum photosynthetic and growth rates in full sunlight. Therefore, any decrease in incident light from canopy shading by weed or crop will decrease rates of growth. However, physiological and morphological acclimation to changing light environments during the growth of weeds can reduce the growth-limiting effects of shading, enabling weeds that have developed in shade to reach high rates of photosynthesis and growth if they are, subsequently, exposed to high light.[48] This capacity for acclimation is an important component of the plasticity characteristic of the "general purpose genotype" typical of many weeds.[31]

Plants grown in shade typically have lower maximum photosynthetic rates than plants grown in full sunlight. However, because of acclimation, their photosynthetic rates at low light generally are greater than those of plants grown in full sunlight. Shade-grown plants normally have lower rates of dark respiration which result in conservation of the limited photosynthate produced. As a result of the lower respiration rates, shade-grown plants typically have a lower light compensation point, the light level at which respiratory CO_2 loss equals photosynthetic CO_2 uptake, and no net CO_2 exchange occurs. Shading also tends to decrease the activity of $RUBP_c$ and PEP_c as well as the leaf chlorophyll content per unit area. The electron transport capacity also may be reduced.

In C_4 plants, growth in dense shade may disrupt the specialized "Kranz" anatomy[78] involved in the functioning of the C_4 pathway in normal leaves. For example, in *Rottboellia exaltata* L.f. grown at 2 or 25% of full sunlight, chloroplasts in the mesophyll cells appeared to migrate away from the mesophyll-bundlesheath cell boundary.[79] Shading typically decreases leaf mesophyll thickness and reduces the ratio of internal-to-external leaf area or A^{mes}/A.[80] Decreases in A^{mes}/A cause increases in the apparent mesophyll resistance to CO_2 uptake in shaded plants and are one reason why shade-grown plants have lower maximum photosynthetic rates than sun-grown plants.[81]

3. *Adaptation to Changing Light Environments*

Changes in the physiological and morphological characteristics of leaves may occur rapidly when plants are exposed to different light environments. In corn and *Amaranthus palmeri* S. Wats., acclimation of photosynthetic rates and PEP_c activities occurred within 6 days after transfer from low to high light or vice versa.[82] Exposure of shade-grown plants of *Solanum dulcamara* L. to high light for 1 to 6 days caused increases in the thickness of leaves which had fully expanded under low light.[83] Chlorophyll and soluble protein contents per unit leaf area, $RUBP_c$ activity, and light-saturated photosynthetic rate also increased rapidly after plants were transferred from low to high light. When *Eichornia crassipes* (Mart. Solms) plants grown for 120 days at low light (90 $\mu E\ m^{-2}\ s^{-1}$) were transferred to high light (750 $\mu E\ m^{-2}\ s^{-1}$) for 5 days, the light-saturated photosynthetic rates of young leaves increased by 45%.[84] This increased photosynthetic capacity was associated with increases in soluble protein content per unit leaf area, specific leaf weight, and mesophyll conductance. In a study of soybean grown at 250 or 850 $\mu E\ m^{-2}\ s^{-1}$ and transferred from one light level to the other, Bunce et al.[85] found that changes in photosynthetic rate, leaf thickness, A^{mes}/A, and stomatal conductance occurred within 1 day.

At the whole-plant level, adaptation to shade occurs with increases in the leaf area ratio (LAR) or the amount of light-harvesting assimilatory surface per unit of plant biomass.[86] Increases in LAR in shaded plants can result from increases in both the leaf weight ratio (LWR) and the specific leaf area (SLA) since LAR = LWR × SLA. Increases in LWR reflect a greater partitioning of plant biomass into the leaf component, while increases in SLA indicate that each unit of leaf biomass is being distributed as a greater amount of leaf area. Significant alterations in the LAR can occur when the light environment in which a plant grows is changed. For example, when *Cyperus rotundus* L. was grown for 30 days in full sunlight and transferred to 15% sunlight, the leaf area ratio, after 32 days in the shade treatment, increased by 38%. In the reciprocal transfer (15% sunlight for 30 days, then full sunlight for 32 days) the leaf area ratio decreased by 30%.[87] Similar morphological and physiological adjustments or acclimation in response to changes in the light environment have been reported for a number of other weeds including *C. esculentus* L.,[87] *Crotalaria spectabilis* Roth,[88] and *Imperata cylindrica* (L.) Beauv.[89]

Moderate shading may stimulate height growth in weeds. In *Rottboellia exaltata* L.f., growth in 60% sunlight for 40 days produced taller plants than growth in full sunlight.[90] Mercado[91] also reported that shading increased height growth in *Rottboellia*. These responses to shading help explain the competitiveness of *Rottboellia* with corn. Thomas and Allison[92] reported that *Rottboellia* competed with corn for light rather than for soil nutrients or water.

B. Temperature

Temperature is an environmental variable of major importance in limiting the geographic distribution of weeds and crops, determining the time of year during which growth and reproductive development may occur, and controlling the rates of plant growth and development. Most physiological processes exhibit upper temperature limits and lower temperature thresholds, above and below which the process will not occur. The optimum temperature, at which the process reaches its maximum rate, lies somewhere between the lower threshold and upper-limit temperature. These three temperatures (lower threshold, optimum, and upper limit) constitute the cardinal temperatures.[93]

1. Short- and Long-Term Effects

In influencing plant growth and physiological responses, temperature acts primarily through its effects on enzymes that catalyze limiting chemical reactions. As temperature increases, the reaction rate increases, but increased enzyme degradation also occurs. The temperature response curve of a process, thus, reflects the sum total of these trends, reaction rate increase, and enzyme degradation.[93] Temperature also may act through its influence on the physical properties of water. Thus, reduced absorption of water by plant roots in cold soil can be partly attributed to the increased viscosity of water.[94]

The temperature to which a plant is exposed may have immediate effects such as those related to enzymatic catalysis of chemical reactions or long-term delayed effects. These delayed or inductive effects exert control over many ontogenetic and morphogenetic processes including the breaking of seed dormancy (stratification) and the stimulation of floral development in biennial and winter annual weeds (vernalization).[93]

Temperature influences photosynthetic CO_2 uptake through both short- and long-term effects.[48] Long-term temperature-related anatomical changes that may affect photosynthesis include changes in stomatal density per unit leaf area with consequent changes in leaf stomatal diffusive resistance. Changes in leaf mesophyll thickness or volume per unit area also occur, resulting in effects on mesophyll conductance. Leaf chlorophyll content and size and density of photosynthetic units also may vary with temperature during growth. Short-term temperature effects include changes in the activity of carboxylation enzymes, modification of electron transport capacity, changes in Hill reaction activity, variations in the rate of CO_2

evolution through photorespiration, and alteration of rates of synthesis, degradation, and translocation of carbohydrates.

2. Acclimation and Adaptation to Changing Temperatures

Trends and changes in the above processes are reflected in the phenomenon of photosynthetic temperature acclimation. This acclimation shifts the temperature optimum for photosynthesis toward the direction of the temperature at which the plant grows.[48] Changes in the upper and lower cardinal temperatures also can occur during acclimation. Capacities for temperature acclimation vary among species and among populations within a single species. For example, Pearcy[95] reported greater capacities for photosynthetic acclimation to high temperatures in desert ecotypes of *Atriplex lentiformis* (Torr.) Wats. in comparison with ecotypes from a mild, coastal environment. In *Oxyria digyna* (L.) Hill, plants from alpine populations which are normally subjected to a widely fluctuating diurnal temperature range had a greater capacity for temperature acclimation than plants from arctic populations.[96] In 15 of 17 populations studied, plants grown at higher temperatures had higher temperature optima for net photosynthesis and higher upper temperature limits than comparable plants grown at lower temperatures. The upper temperature limit or compensation point for photosynthesis occurs when respiratory CO_2 loss reaches a level that results in no net photosynthetic CO_2 uptake.

The capacity for photosynthetic temperature acclimation may increase the competitiveness of a species. For example, the competitive success of winter annual species in the early stages of old field succession was attributed to their ability to maintain relatively high rates of net photosynthesis during the winter and early spring, despite prevailing low temperature.[97] In the short-grass prairie of the western U.S., the C_3 grass *Agropyron smithii* Rydb. lacks the ability to acclimate to high temperatures, while its potential competitor, the C_4 grass *Bouteloua gracilis* (H.B.K.) Lag. ex Steud., cannot acclimate to low temperatures. This difference in acclimation capacity is one factor accounting for the temporal separation of the active growth periods of these grasses.[98]

3. Effects on Growth and Competitiveness

Long-term effects of temperature on plant growth and competitiveness may be evaluated using the techniques of mathematical growth analysis. With these techniques the relative importance of the leaf area factor can be demonstrated. In a comparison of corn, cotton, soybean, and six weed species, Potter and Jones[99] reported that relative growth rates and relative leaf area expansion rates were greater in the weeds than in the crops. Growth in all nine species was greater at day/night temperatures of 32/21°C than at 38/27 or 21/10°C. The relative growth rate (R_w) was highly correlated with the leaf-area partitioning coefficient, a measure of the proportion of the daily biomass increment that is converted to leaf area. In contrast, R_w was poorly correlated with NAR. Similar conclusions were reached in studies of temperature effects on fescue.[100] In *C. spectabilis* Roth, variations in dry matter production at day/night temperatures of 23/17, 26/20, 29/23, and 32/26°C were highly positively correlated with variations in leaf area duration (LAD) (r^2 = 0.988). On the other hand, dry matter production was negatively correlated with NAR.[91] These results from growth analysis experiments emphasize the need to consider changes in the leaf area component when temperature responses are studied.

Comparative studies of the temperature responses of weeds and crops may help explain their competitive interactions in the field under different environmental conditions. For example, field observations in the southern U.S. indicate that the malvaceous weeds *Abutilon theophrasti* Medic. and *Anoda cristata* (L.) Schlecht. are more competitive with cotton following periods of abnormally cool temperatures early in the growing season.[101,102] Experiments in controlled-environment chambers showed that the weeds were less sensitive than cotton to simulated short-term early-season chilling events.[103] Leaf area expansion rates, which were reduced by chilling in all three species, recovered more rapidly in *Abutilon* and *Anoda*. As a result of these differential effects of chilling, the weeds gained a greater advantage over cotton in leaf area expansion and dry matter production. Chilling significantly reduced stomatal conductance and net photosynthesis in both cotton and *Abutilon*, but recovery to the levels of unchilled controls was complete within 94 h. Thus, the primary effect of chilling was to reduce the rates of leaf initiation and leaf expansion. Subsequent competition experiments with the same three species confirmed that the relative competitiveness of cotton was decreased by cool temperatures.[104]

Competitive differences in C_3 and C_4 plants have been demonstrated in studies conducted under temperature conditions known to favor the growth of one or the other type. For example, Pearcy et al.[105] compared photosynthesis, growth, and competitiveness of C_3 *Chenopodium album* L. and C_4 *Amaranthus retroflexus* L. in three day/night temperature regimes. Growth at 17/14°C reduced the net photosynthetic rate per unit leaf area of *Amaranthus* below that of *Chenopodium;* growth at 25/18°C resulted in similar photosynthetic rates in both species; and growth at 34/28°C resulted in a significantly higher net photosynthetic rate in *Amaranthus* than in *Chenopodium*. Similar trends in relative growth rates and height growth reflected the influence of variations in photosynthetic rate on these parameters. Analysis of competitive interactions between the two species in the three temperature regimes revealed a clear competitive advantage for C_3 *Chenopodium* at 17/14°C and for C_4 *Amaranthus* at 34/28°C. Pearcy et al.[105] concluded that species differences in relative growth rate and height growth responses to temperature, before canopy closure, were primarily responsible for the observed differences in competitive ability between the C_3 and C_4 plants. In similar competition experiments with the prairie grasses *Agropyron smithii* Rydb. and *B. curtipendula* (Michx.) Torr., Christie and Detling[106] found that the competitiveness of C_3 *Agropyron* was favored by low temperatures of 20/12°C, while competitiveness of C_4 *Bouteloua* was greater at 30/15°C.

Flint and Patterson[107] investigated the effects of temperature on the growth and competitiveness of the C_3 crop soybean, the C_3 weed *Xanthium pensylvanicum* Wallr., and the C_4 weed *Amaranthus hybridus* L. At 34 days after planting, height growth, dry matter production, and leaf area expansion increased with increasing day/night temperature from 26/17 to 32/23°C in all three species, but the greatest increase with temperature occurred in *Amaranthus*. Analysis of relative yield values of the species grown in competition showed that *Xanthium* and soybean were approximately equivalent competitors with each other regardless of temperature. The competitiveness of *Amaranthus* with the C_3 species was greatly reduced at the lowest temperature. However, regardless of temperature, the two C_3 species were superior competitors to *Amaranthus*. This may be attributed to the greater seed size and resulting greater seedling size of the C_3 species compared to *Amaranthus* (soybean 190 mg per seed, *Xanthium* 26.6 mg per seed, *Amaranthus* 0.4 mg per seed). Greater seed size has been shown to confer a competitive advantage during early growth in a number of species.[38,108] Many factors other than maximum photosynthetic rate contribute to competitive ability in both weeds and crops. For example, in the above study, the weakest competitor, *Amaranthus*, had the greatest net assimilation rate. This was to be expected since it is a C_4 plant. However, competitiveness was more closely related to leaf area duration which was much greater in the two C_3 species.

Studies of their temperature responses also are useful in predicting the potential geographic distribution and agronomic impact of recently introduced exotic weeds. *Rottboellia exaltata* L.f., a weed of Old World tropical origin, presently is confined in the U.S. to the Gulf Coast states. Experimental studies of *Rottboellia* in 36 day/night temperature combinations ranging from 17/11 to 32/26°C demonstrated that the potential range of the weed in the U.S. is much greater than its present range.[109] *Rottboellia* can flower and produce seed in temperature regimes simulating the warmest 4 months of the growing season as far north as Wisconsin. However, superimposed simulated early-season chilling events reduced the growth of *Rottboellia* more than that of the regionally adapted varieties of corn and soybean with which it would be expected to compete in regions outside the extreme southern U.S.[110] The studies indicate that *Rottboellia* is unlikely to be an effective early-season competitor with corn or soybean. However, other studies have shown *Rottboellia* to have some shade tolerance.[88,89] Therefore, *Rottboellia* might persist under a crop canopy until warmer temperatures occur later in the growing season. Its rapid growth rate and profuse tillering under warmer conditions might enable *Rottboellia* to compete more effectively with the established crop.

C. Water

In many agricultural situations, water availability is the major factor limiting crop productivity.[94,111] Weeds compete with crops for the often limited soil water resource and further reduce the availability of water to the crop. Reduced soil water availability increases the level of water stress experienced by the crops, because water stress develops in plants when transpirational water losses from the leaves exceed water uptake from the soil through the roots.

1. Effects of Water Stress on Physiological Processes and Growth

The effects of water stress on plant growth and physiological processes have been reviewed extensively.[94,112,113] The effects of water stress vary with the severity and duration of the water deficit, the environmental conditions under which the plant has grown, the species of plant experiencing the stress, the growth stage at which water stress occurs, and the physiological processes considered.

Water stress reduces photosynthesis by causing stomatal closure through reduced turgor in the guard cells. Stomatal closure, in turn, reduces the rate of diffusion of atmospheric CO_2 into the sites of initial carboxylation in the leaf mesophyll. Water stress also may directly affect various components of the photosynthetic process, including chlorophyll synthesis in greening leaves,[114] electron transport and photophosphorylation,[115] and the activity of carboxylating enzymes.[116] Photorespiration tends to decrease with water stress, while dark respiration is less altered.[117]

Water stress reduces the total photosynthetic capacity of a plant by reducing the rate of leaf area expansion and, consequently, the total amount of photosynthetic assimilatory surface present. Leaf expansion is one of the processes most sensitive to water stress, and reductions in rates of leaf expansion often occur before other indications of water stress become apparent.[118]

2. Acclimation to Water Stress

Acclimation to water stress through the process of osmotic adjustment can reduce the growth-limiting effects of water stress.[49] Osmotic adjustment in plant tissue can occur through synthesis or absorption of solutes (osmo-regulation) or through dehydration which effectively increases the concentration of solutes already present. The process of osmotic adjustment decreases the solute potential of water in the plant tissue. In the leaf, osmotic adjustment maintains the positive turgor required for leaf expansion, under conditions of decreasing total leaf water potential. In stomatal guard cells, osmotic adjustment also results in maintenance of turgor and, hence, stomatal opening, under lower leaf water potentials. This enables CO_2 absorption and photosynthesis to continue at lower leaf water potentials.

3. Reasons for Species' Differences in Tolerance of Water Stress

Different species of weeds and crops vary widely in their responses to low soil water availability and plant water stress. For example, Wiese and van Diver[119] reported that the relative competitive abilities of corn, sorghum, and eight species of weeds varied with soil moisture level. With high soil moisture levels, weeds typical of humid regions or normally present in irrigated crops were most competitive. Conversely, weeds native to arid regions were more competitive when soil moisture was more limited.

Such interspecific differences in response to water availability may reflect differences in root distribution and rate of root elongation, in the tolerance of low water potential in various plant tissues, and in the control of transpirational water loss.

The patterns of root distribution in various weeds and crops have been examined by carefully excavating the root systems[120,121] or by employing other, indirect measures such as determining the pattern of soil water extraction around isolated plants.[122] Davis et al.[122] reported root profiles for sorghum and seven species of weeds grown in field plots without competition for 5 months. *Xanthium pensylvanicum* Wallr. had by far the most extensive root system with a maximum depth of 2.9 m and a lateral spread of 8.6 m. *Kochia scoparia* (L.) Schrad., *Digitaria sanguinalis* (L.) Scop., and *Tribulus terrestris* L. also had large root systems. The root system of sorghum was considerably less extensive with a depth of 1.7 m and a lateral spread of 4.3 m. *Salsola iberica* Sennen et Pau, *Amaranthus retroflexus* L., and *Solanum rostratum* Dunal also had relatively small root systems. However, the root system of *Amaranthus* developed very rapidly, reaching 88% of its maximum extent after only 10 weeks. The rapid root growth of *Amaranthus* contributed to its ability to compete effectively with sorghum.

4. Water Use Efficiency

Rates of water loss by weed and crop plants grown in large pots under field conditions were studied extensively early in this century.[123,124] These studies revealed much lower rates of water loss and greater water use efficiencies in plants now known to have the C_4 photosynthetic pathway. The "water requirement", or amount of water lost per unit of dry matter produced, ranged from 1068 in the C_3 crop plant alfalfa (*Medicago sativa* L.) to 303 in the C_4 weed pigweed (*Amaranthus* spp.).

More recently, Patterson and Flint[125] compared single-leaf and whole-plant water loss rates of soybean and seven species of weeds under controlled-environment conditions. Water use efficiencies of whole plants (milligrams dry matter produced per gram of water lost) and photosynthesis/transpiration ratios (milligrams CO_2 fixed per gram of water lost) also were calculated. Among the eight species, *Amaranthus hybridus* L., the only C_4 plant, had the highest net photosynthetic rate and net assimilation rate and the lowest transpiration rate on a single-leaf or whole-plant basis (Table 6). Water use efficiencies and photosynthesis/transpiration ratios of *Amaranthus* were two- to threefold greater than those of the other species. As mentioned previously, the rate of photosynthetic CO_2 uptake in C_4 plants is less limited by stomatal resistance than is the case with C_3 plants, because of the CO_2-concentrating function of the C_4 pathway. Thus, even with a high stomatal resistance which greatly reduces transpiration, *Amaranthus* can maintain a high net photosynthetic rate. Among the seven C_3 species, soybean and *Xanthium pensylvanicum* Wallr. had the greatest rates of water loss per plant. However, these high rates resulted from greater amounts of leaf area per plant rather than from greater transpiration rates per unit leaf area.

Examination of the relationship between stomatal resistance and decreasing leaf water potential (LWP) as water was withheld from the potted plants revealed several different patterns among the eight species. In *Datura stramonium* L., stomatal closure began at a LWP of about −9 bars, whereas in *Sida spinosa* L. the stomata remained open until a LWP of −16 to −20 bars occurred. In *Anoda cristata* (L.) Schlecht and *Amaranthus hybridus* L., stomatal closure occurred more gradually as LWP declined from −8 to −20 bars. These differences in stomatal response to water stress may be significant in determining the impact

Table 6
NET PHOTOSYNTHETIC RATES, NET ASSIMILATION RATES, SINGLE-LEAF AND WHOLE-PLANT TRANSPIRATION RATES, PHOTOSYNTHESIS/TRANSPIRATION RATIOS, AND WATER USE EFFICIENCIES OF SOYBEAN AND SEVEN SPECIES OF WEEDS[a]

Species	Net photosynthetic rate (mg CO_2 dm^{-2} h^{-1})	Net assimilation rate (g dm^{-2} day^{-1})	Single-leaf transpiration rate (g H_2O dm^{-2} h^{-1})	Whole-plant transpiration rate (g H_2O dm^{-2} day^{-1})	Photosynthesis transpiration ratio (mg CO_2/g H_2O)	Water use efficiency (mg dry biomass/g H_2O)
Glycine max (L.) Merr.	41	0.11	11.4	28.0	3.6	4.0
Abutilon theophrasti Medic.	42	0.12	8.9	29.5	4.7	4.3
Amaranthus hybridus L.	96	0.17	6.7	20.6	14.6	8.7
Anoda cristata (L.) Schlect.	52	0.13	12.8	33.1	4.2	4.0
Cassia obtusifolia L.	47	0.11	10.7	27.1	4.4	4.0
Datura stramonium L.	53	0.15	13.0	36.5	4.2	4.4
Sida spinosa L.	50	0.14	14.2	29.4	3.8	4.8
Xanthium pensylvanicum Wallr.	50	0.13	13.4	30.5	3.9	4.4

[a] Data collected in controlled-environment greenhouse with natural lighting and day/night temperature of 32/23°C.

From Patterson, D. T. and Flint, E. P., *Weed Sci.*, 31, 318, 1983. With permission.

of weed competition on water availability. Species such as *S. spinosa* L., which have relatively high transpiration rates and maintain open stomata under increasing water stress, can be considered exploitative "water users", whereas species with stomata that are more responsive to declining LWP are "water savers". In a competitive situation in which soil moisture is the primary limiting resource, the "water user" would usurp the resource to the detriment of the "water saver".[126] Further studies of weed/crop competition for water should take this theory into account.

D. Nutrients

Together with light and water, soil mineral nutrients comprise the major environmental resources for which weeds and crops compete.[36] Depending on the relative supply of any of these resources and the degree of plant cover, competition may be more or less severe.

1. Long-Term Effects of Soil Fertility Differences

Long-term studies have demonstrated substantial effects of soil-fertility treatments on the weed flora in various crops. In permanent grassland plots in England, supplemental nitrogen favored the predominance of grasses over legumes, especially when ammonium sulfate, which increases soil acidity, was the nitrogen source.[127] In land in continuous winter wheat production for 47 years, weed seed populations in the soil were greatest in plots receiving full nutrient applications.[128] High fertility treatments favored grasses and sedges at the expense of broad-leaved weeds.

2. Competition for Nutrients

Among the major plant nutrients, nitrogen is the one for which competition is most severe.[129] However, interactions between competition for different nutrients as well as for light and water may exist to the extent that increasing the level of one resource increases competition for a second resource. Thus, the yield-reducing effects of weeds, even when demonstrably caused by competition for nutrients, may not be eliminated by the addition of fertilizer. For example, when supplemental nitrogen fertilizer was applied to fields of upland rice infested with *Cyperus rotundus* L., the weed benefited more than the rice. This occurred because *Cyperus* competed more effectively for soil moisture under high nitrogen conditions.[130] On the other hand, adding high levels of nitrogen to cotton in competition with annual weeds resulted in a substantial increase in the height of the cotton plants. This gave cotton an advantage in competition for light not realized at low nitrogen levels.[131]

Weed/crop interactions and competition for soil nutrients are further complicated by the presence of microorganisms. For example, *Cyperus esculentus* L. grown in association with silage corn reduced the dry matter yield and total nitrogen content of the crop.[132] However, only 38% of the decrease in the nitrogen content of the crop could be accounted for through uptake by the weed. Further study indicated that the presence of *Cyperus* roots caused increased activity of denitrifying bacteria with a consequent decrease in available nitrogen. Further examples of such complex interactions are considered in the chapter on allelopathy.

Evaluation of the growth analysis parameters NAR and LAR, whose product (Table 1) is approximately equal to the relative growth rate (R_w), can be useful in assessing the effects of nutrient deficiencies arising from weed/crop competition. According to Blackman,[19] nitrogen deficiency reduces LAR, while with the exception of extreme N deficiencies, NAR is unaffected. Potassium deficiency, on the other hand, reduces NAR more than LAR. Phosphorus deficiency reduces both NAR and LAR. Deficiencies in the two other major resources for which plants compete, i.e., water and light, also affect R_w.[19] Water deficits tend to decrease NAR more than LAR. Shading, or light deficiency, increases LAR and decreases NAR. The relative changes in LAR and NAR in response to shading determine whether a particular plant will reach its optimum growth in full sunlight (true for most weeds and crops) or at some lower light level (the case for obligate shade plants, such as some forest floor herbs). Consideration of the relationships described above should facilitate the analysis of the complex interactions involved in weed/crop competition for nutrients and other resources.

E. Carbon Dioxide

1. Competition for CO_2

Although CO_2 is an essential resource for plant growth, there is some question whether competition for CO_2 occurs in nature.[133] By definition, competition occurs only when two or more organisms strive for the same limited resource so that the presence of one organism reduces the availability of the resource to the other organism.[24] However, in a rapidly photosynthesizing weed or crop canopy, the CO_2 concentration may be depressed below normal ambient levels. In this situation, competition for CO_2 would occur if the growth of one plant were depressed as a result of the depletion of the CO_2 resource by a neighboring plant of the same or different species. Competition for CO_2 should be most intense in mixtures of C_3 and C_4 plants, because of their differing ability to withdraw CO_2 from the air through photosynthesis. The absence of a CO_2 concentrating mechanism in C_3 plants results in a greater reduction of their photosynthetic rates by low CO_2 than is the case with C_4 plants.

2. Potential Effects of Increased Atmospheric CO_2 Concentration

Our concern with CO_2 results also from the observation that its concentration in the global atmosphere is increasing, and various species of plants react differently to increased CO_2.[134] Current projections indicate a possible doubling of the atmospheric CO_2 concentration from the present 330 ppm within the next 100 years. Because of the so-called greenhouse effect, a doubling of the atmospheric CO_2 concentration may result in a significant increase in the mean global temperature, with consequent increased aridity over much of the earth's surface.[135] Such indirect, climatic effects as increased temperature and aridity may influence the distribution and productivity of major crops and their interaction with weeds and other crop pests.

Various studies have demonstrated a "fertilization" effect of CO_2, an enhancement of plant growth at CO_2 concentrations above ambient levels.[136] This phenomenon is particularly well known in horticultural crops produced in greenhouses where artificial enrichment of the CO_2 is feasible. However, all plants do not respond to CO_2 enrichment in the same ways.

Of particular relevance to weed/crop interactions is the observation that photosynthetic and growth rates of C_3 plants are stimulated more by CO_2 enrichment than those of C_4 plants. As mentioned earlier, the greater stimulation of photosynthesis in C_3 plants occurs because they lack the specialized leaf anatomy and metabolic pathway that concentrate atmospheric CO_2 at the site of fixation by $RUBP_c$ in C_4 plants. Differential stomatal responses to enriched CO_2 concentrations also have been reported in C_3 and C_4 plants.[137] As we have noted earlier, most major agronomic crops are C_3 plants, whereas many of their competing weeds are C_4 plants. Therefore, any factor such as CO_2 enrichment which affects C_3 and C_4 plants differently may be expected to influence weed/crop interactions.

Differential effects of CO_2 enrichment on C_4 and C_3 weeds and crops have been demonstrated in controlled-environment chambers. Increasing the CO_2 concentration from 350 to 600 or 1000 ppm increased the net assimilation rates of the C_3 weed *Abutilon theophrasti* Medic. and the C_3 crop soybean.[138] However, in the C_4 weed *Rottboellia exaltata* L.f. and the C_4 crop corn, net assimilation rates did not increase. As a result, dry matter production was stimulated to a greater extent by CO_2 enrichment in the C_3 species. In other work, increasing the CO_2 concentration from 350 to 675 ppm increased net assimilation rate, leaf area duration, and dry matter production more in soybean than in the C_4 weed *Sorghum halepense* (L.) Pers.[139] When the two species were grown together in mixed culture, the plant relative yield (PRY) of soybean in competition with *S. halepense* (L.) Pers. increased, and the PRY of *S. halepense* (L.) Pers. decreased at 675 ppm CO_2 as compared with 350 ppm CO_2. Thus, the competitiveness of the C_3 crop with the C_4 weed was enhanced by CO_2 enrichment.

Whether such differential effects of CO_2 enrichment will occur under field conditions remains a question, especially when we consider that water stress and nutrient deficiency, rather than atmospheric CO_2 concentration, often limit plant growth.[140] In addition, the potential for genetic adaptation and phenotypic adjustment among weeds and crops in response to long-term, gradual increases in atmospheric CO_2 concentration is unknown. However, recent controlled environment studies have shown that plant growth enhancement by CO_2 enrichment is not eliminated by water stress or limited nutrient availability.[141,142] Therefore, the likelihood that future increases in the global atmosphere CO_2 concentration will significantly affect weed/crop interactions must be considered.

V. SUMMARY

Crops and associated weeds share many characteristics, including, in some cases, evolutionary history and taxonomic classification into relatively few plant families. However, agronomic weeds have retained greater phenotypic plasticity than most intensively selected crops. This plasticity includes the capacity for reversible and irreversible physiological and morphological changes which occur in response to variations in the environment in which the weeds grow. This short- and long-term adaptation to environmental variables confers some degree of freedom from environmental constraints.

Many of the most pernicious weeds have the C_4 photosynthetic pathway, which, in contrast, occurs in only a few of the major crops of the world. The C_4 pathway appears to be of adaptive value in environments characterized by high light, high temperature, and limited moisture. The relative proportion of C_4 vs. C_3 weeds infesting major crops in the U.S. is related to the time of year during which the crop is grown, rather than to geographic region, warm season crops having a high proportion of C_4 weeds.

Competition for water, nutrients, and light is the primary cause of the losses in crop yield associated with weed infestation. The intensity of competition for these resources varies with the scarcity of the resource and the relative demand for the resource exerted by crop and weed. World-wide, the availability of water is the primary limitation to crop productivity. Weeds reduce the availability of water to the crop and this can result in greater levels of water stress in the crop. In many crop/weed associations, competition for nitrogen is a major cause of crop yield reductions. However, supplemental fertilization may stimulate the growth of the weeds more than that of the crop, to the extent that competition for light or water becomes more severe.

A better knowledge of weed/crop ecophysiology can help us understand the complex responses and interactions that occur in the agroecosystem.

REFERENCES

1. **Holm, L. G., Pancho, J. V., Herberger, J. P., and Plucknett, D. L.,** *A Geographical Atlas of World Weeds*, John Wiley & Sons, New York, 1979.
2. **Holm, L. G., Plucknett, D. L., Pancho, J. V., and Herberger, J. P.,** *The World's Worst Weeds*, University Press of Hawaii, Honolulu, 1977.
3. **Harlan, J. R.,** *Crops and Man*, American Society of Agronomy, Madison, Wis., 1975.

4. **Evans, L. T.,** The physiological basis of crop yield, in *Crop Physiology, Some Case Histories*, Evans, L. T., Ed., Cambridge University Press, Cambridge, 1975, chap. 11.
5. **Kramer, P. J.,** The role of physiology in crop improvement, in *Linking Research to Crop Production*, Staples, R. C. and Kuhr, R. J., Eds., Plenum Press, New York, 1980, 51.
6. **Patterson, D. T.,** Effects of light and temperature on weed/crop growth and competition, in *Biometeorology in Integrated Pest Management*, Hatfield, J. L. and Thomason, I. J., Eds., Academic Press, New York, 1982, 407.
7. **Huffaker, C. B.,** Fundamentals of biological weed control, in *Biological Control of Insect Pests and Weeds*, DeBach, P., Ed., Reinhold, New York, 1964, chap. 22.
8. **Patterson, D. T.,** Phenotypic and physiological comparisons of field and phytotron grown plants, in *Crop Reactions to Water and Temperature Stresses in Humid, Temperate Climates*, Raper, C. D., Jr. and Kramer, P. J., Eds., Westview Press, Boulder, Colo., 1983, chap. 19.
9. **Evans, L. T.,** Extrapolation from controlled environments to the field, in *Environmental Control of Plant Growth*, Evans, L. T., Ed., Academic Press, New York, 1963, 421.
10. **Kramer, P. J.,** The use of controlled environments in research, *HortScience*, 13, 447, 1978.
11. **Janac, J., Catsky, J., and Jarvis, P. G.,** Infrared gas analyzers and other physical analyzers, in *Plant Photosynthetic Production, Manual of Methods*, Sestak, Z., Catsky, J., and Jarvis, P. G., Eds., W. Junk Publishers, The Hague, 1971, 11.
12. **Marshall, B. and Biscoe, P. V.,** A mobile apparatus for measuring leaf photosynthesis in the field, *J. Exp. Bot.*, 28, 1008, 1977.
13. **Kanemasu, E. T., Thurtle, G. W., and Tanner, C. B.,** Design, calibration, and use of a stomatal diffusion porometer, *Plant Physiol.*, 44, 881, 1969.
14. **Ritchie, G. A. and Hinckley, T. M.,** The pressure chamber as an instrument for ecological research, *Adv. Ecol. Res.*, 9, 165, 1975.
15. **Slavik, B.,** *Methods for Studying Plant Water Relations*, Springer-Verlag, New York, 1974, 38.
16. **Hanson, A. D. and Hitz, W. D.,** Metabolic responses of mesophytes to plant water deficits, *Annu. Rev. Plant Physiol.*, 33, 163, 1982.
17. **Bunce, J. A.,** Leaf elongation in relation to leaf water potential in soybean, *J. Exp. Bot.*, 28, 156, 1977.
18. **Kvet, J., Ondok, J. P., Necas, J., and Jarvis, P. G.,** Methods of growth analysis, in *Plant Photosynthetic Production, Manual of Methods*, Sestak, Z., Catsky, J., and Jarvis, P. G., Eds., W. Junk Publishers, The Hague, 1971, 343.
19. **Blackman, G. E.,** The application of the concepts of growth analysis to the assessment of productivity, in *Functioning of Terrestrial Ecosystems at the Primary Production Level*, Eckardt, F. E., Ed., UNESCO, Paris, 1968, 243.
20. **Evans, G. C.,** *The Quantitative Analysis of Plant Growth*, University of California Press, Berkeley, 1972.
21. **Watson, D. J.,** The physiological basis of variation in yield, *Adv. Agron.*, 4, 101, 1952.
22. **Harper, J. L.,** *Population Biology of Plants*, Academic Press, London, 1977, chap. 8.
23. **Trenbath, B. R.,** Biomass productivity of mixtures, *Adv. Agron.*, 26, 177, 1974.
24. **Donald, C. M.,** Competition among crop and pasture plants, *Adv. Agron.*, 15, 1, 1963.
25. **Remison, S. U. and Snaydon, R. W.,** A comparison of root competition and shoot competition between *Dactylis glomerata* and *Holcus lanatus*, *Grass Forage Sci.*, 35, 183, 1980.
26. **Anderson, E.,** *Plants, Man and Life*, University of California Press, Berkeley, 1971, chap. 2, 10.
27. **Harlan, J. R. and deWet, J. M. J.,** Some thoughts about weeds, *Econ. Bot.*, 19, 16, 1965.
28. **Harlan, J. R.,** The possible role of weed races in the evolution of cultivated plants, *Euphytica*, 14, 173, 1965.
29. **Stebbins, G. L.,** *Flowering Plants, Evolution Above the Species Level*, Harvard University Press, Cambridge, 1974, 355.
30. **Baker, H. G.,** Characteristics and modes of origin of weeds, in *The Genetics of Colonizing Species*, Baker, H. G. and Stebbins, G. L., Eds., Academic Press, New York, 1965, 147.
31. **Baker, H. G.,** The evolution of weeds, *Annu. Rev. Ecol. Syst.*, 5, 1, 1974.
32. **Hill, T. A.,** *The Biology of Weeds*, Edward Arnold, London, 1977, chap. 3, 4.
33. **Sharp, D.,** A phytosociological study of weed communities on the southwestern coastal plain of North Carolina, *Vegetatio*, 31, 103, 1976.
34. **Grime, J. P.,** *Plant Strategies & Vegetation Processes*, John Wiley & Sons, Chichester, 1979, chap. 1 and 2.
35. **Barrett, S. C. H.,** Crop mimicry in weeds, *Econ. Bot.*, 37, 255, 1983.
36. **Zimdahl, R. L.,** *Weed-Crop Competition, A Review*, International Plant Protection Center, Corvallis, Ore., 1980, 83.
37. **Harper, J. L.,** The nature and consequences of interference amongst plants, in *Genetics Today*, Vol. 2, Geerts, S. J., Ed., Pergamon Press, Oxford, 1965, 465.
38. **Clements, F. E., Weaver, J. E., and Hanson, H. C.,** *Plant Competition — An Analysis of Community Function*, Carnegie Institute, Washington, D.C., 1929.

39. **Bunting, A. H.,** Some reflections on the ecology of weeds, in *The Biology of Weeds*, Harper, J. L., Ed., Blackwell Scientific, Oxford, 1960, 14.
40. **Johnston, S. K., Murray, D. S., and Williams, J. C.,** Germination and emergence of balloonvine *(Cardiospermum halicacabum), Weed Sci.*, 27, 73, 1979.
41. **Harlan, H. V.,** The weedishness of wild oats, *J. Hered.*, 20, 515, 1929.
42. **Baker, H. G.,** Human influences on plant evolution, *Econ. Bot.*, 26, 32, 1972.
43. **Panetsos, C. and Baker, H. G.,** The origin of variation in "wild" *Raphanus sativus* (Cruciferae) in California, *Genetica (The Hague)*, 38, 243, 1967.
44. **McNeil, J.,** The taxonomy and evolution of weeds, *Weed Res.*, 16, 399, 1976.
45. **Weaver, S. E. and Lechowicz, M. J.,** The biology of Canadian weeds. *Xanthium strumarium* L., *Can. J. Plant Sci.*, 63, 211, 1983.
46. **Harper, J. L.,** Factors controlling plant numbers, in *The Biology of Weeds*, Harper, J. L., Ed., Blackwell Scientific, London, 1960, 119.
47. **Billings, W. D.,** Adaptations and origins of alpine plants, *Arct. Alp. Res.*, 6, 129, 1974.
48. **Patterson, D. T.,** Light and temperature adaptation, in *Predicting Photosynthesis for Ecosystem Models*, Vol. 1, Hesketh, J. D. and Jones, J. W., Eds., CRC Press, Boca Raton, Fla., 1980, chap. 8.
49. **McMichael, B. L.,** Water stress adaptation, in *Predicting Photosynthesis for Ecosystem Models*, Vol. 1, Hesketh, J. D. and Jones, J. W., Eds., CRC Press, Boca Raton, Fla., 1980, chap. 7.
50. **Kortschak, H. P., Hartt, C. E., and Burr, G. O.,** Carbon dioxide fixation in sugarcane leaves, *Plant Physiol.*, 40, 209, 1965.
51. **Hatch, M. D. and Slack, C. R.,** Photosynthesis by sugarcane leaves. A new carboxylation reaction and the pathway of sugar formation, *Biochem. J.*, 101, 103, 1966.
52. **Black, C. C., Chen, T. M., and Brown, R. H.,** The biochemical basis for plant competition, *Weed Sci.*, 17, 338, 1969.
53. **Calvin, M. and Benson, A. A.,** The path of carbon in photosynthesis, *Science*, 107, 476, 1948.
54. **Black, C. C.,** Checklists of C_4-dicarboxylic acid photosynthesis plant species, in *CO_2 Metabolism and Plant Productivity*, Burris, R. H. and Black, C. C., Eds., University Park Press, Baltimore, 1976, 425.
55. **Elmore, C. D. and Paul, R. N.,** Composite list of C_4 weeds, *Weed Sci.*, 31, 686, 1983.
56. **McWhorter, C. G. and Patterson, D. T.,** Ecological factors affecting weed competition in soybean, in *World Soybean Research Conference II: Proc.*, Corbin, F. T., Ed., Westview Press, Boulder, Colo., 1979, 371.
57. **Baskin, J. M. and Baskin, C. C.,** A discussion of the growth and competitive ability of C_3 and C_4 plants, *Castanea*, 43, 71, 1978.
58. **Hatch, M. D.,** The C_4 pathway of photosynthesis: mechanism and function, in *CO_2 Metabolism and Plant Productivity*, Burris, R. H. and Black, C. C., Eds., University Park Press, Baltimore, 1976, 59.
59. **Ogren, W. L. and Bowes, G.,** Ribulose diphosphate carboxylase regulates soybean photorespiration, *Nature New Biol. (London)*, 230, 159, 1971.
60. **Campbell, W. H. and Black, C. C.,** The relationship of CO_2 assimilation pathways and photorespiration to the physiological quantum requirement of green plant photosynthesis, *BioSystems*, 10, 253, 1978.
61. **Edwards, G. E., Huber, S. C., Ku, S. B., Rathnam, C. K. M., and Mayne, B. C.,** Variation in photochemical activities of C_4 plants in relation to CO_2 fixation, in *CO_2 Metabolism and Plant Productivity*, Burris, R. H. and Black, C. C., Eds., University Park Press, Baltimore, 1976, 83.
62. **Ehleringer, J. and Bjorkman, O.,** Quantum yields for CO_2 uptake in C_3 and C_4 plants, *Plant Physiol.*, 59, 86, 1977.
63. **Ehleringer, J. R.,** Implications of quantum yield differences on the distributions of C_3 and C_4 grasses, *Oecologia*, 31, 255, 1978.
64. **Bjorkman, O.,** Adaptive and genetic aspects of C_4 photosynthesis, in *CO_2 Metabolism and Plant Productivity*, Burris, R. H. and Black, C. C., Eds., University Park Press, Baltimore, 1976, 287.
65. **Doliner, L. H. and Jolliffe, P. A.,** Ecological evidence concerning the adaptive significance of the C_4 dicarboxylic acid pathway of photosynthesis, *Oecologia*, 38, 23, 1979.
66. **Brown, R. H.,** A difference in N use efficiency in C_3 and C_4 plants and its implications in adaptation and evolution, *Crop Sci.*, 18, 93, 1978.
67. **Schmitt, M. R. and Edwards, G. E.,** Photosynthetic capacity and nitrogen use efficiency of maize, wheat, and rice: a comparison between C_3 and C_4 photosynthesis, *J. Exp. Bot.*, 32, 459, 1981.
68. **Gifford, R. M.,** A comparison of potential photosynthesis, productivity, and yield of plant species with differing photosynthetic metabolism, *Aust. J. Plant Physiol.*, 1, 107, 1974.
69. **Monteith, J. L.,** Reassessment of maximum growth rates for C_3 and C_4 crops, *Exp. Agric.*, 14, 1, 1978.
70. **Waller, S. S. and Lewis, J. K.,** Occurrence of C_3 and C_4 photosynthetic pathways in North American grasses, *J. Range Manage.*, 32, 12, 1979.
71. **Williams, G. J., III and Markley, J. L.,** The photosynthetic pathway type of North American shortgrass prairie species and some ecological implications, *Photosynthetica*, 7, 252, 1973.

72. **Teeri, J. A. and Stowe, L. G.,** Climatic patterns and the distribution of C_4 grasses in North America, *Oecologia*, 23, 1, 1976.
73. **Tieszen, L. L., Senyimba, M. M., Imbamba, S. K., and Troughton, J. H.,** The distribution of C_3 and C_4 grasses and carbon isotope discrimination along an altitudinal and moisture gradient in Kenya, *Oecologia*, 37, 337, 1979.
74. **Eickmeier, W. G.,** Photosynthetic pathway distributions along an aridity gradient in Big Bend National Park, and implications for enhanced resource partitioning, *Photosynthetica*, 12, 290, 1978.
75. USDA, Extent and Cost of Weed Control with Herbicides and an Evaluation of Important Weeds, 1968, A.R.S.-H-1, November 1972; available from the NTIS, Springfield, Va.
76. **Takeda, T., Agata, W., Hakoyama, S., and Tanaka, H.,** Studies on weed vegetation in non-cultivated paddy fields. II. The relation between the ecological distribution of Gramineous C_3 and C_4 weeds and the soil moisture condition in non-cultivated paddy fields, *Jpn. J. Crop Sci.*, 46, 558, 1977.
77. **Smith, H.,** Light quality, photoperception, and plant strategy, *Annu. Rev. Plant Physiol.*, 33, 481, 1982.
78. **Laetsch, W. M.,** The C_4 syndrome: a structural analysis, *Annu. Rev. Plant Physiol.*, 25, 27, 1974.
79. **Paul, R. N. and Patterson, D. T.,** Effects of shading on the anatomy and ultrastructure of the leaf mesophyll and vascular bundles of itchgrass *(Rottboellia exaltata)*, *Weed Sci.*, 28, 216, 1980.
80. **Nobel, P. S., Zarazoza, L. J., and Smith, W. K.,** Relation between mesophyll surface area, photosynthetic rate and illumination level during development of leaves of *Plectranthus parviflorus* Henckel, *Plant Physiol.*, 55, 1067, 1975.
81. **Nobel, P. S.,** Internal leaf area and cellular CO_2 resistance: photosynthetic implications of variations with growth conditions and plant species, *Physiol. Plant.*, 40, 137, 1977.
82. **Hatch, M. D., Slack, C. R., and Bull, T. A.,** Light-induced changes in the content of some enzymes of the C_4-dicarboxylic acid pathway of photosynthetsis and its effect on other characteristics of photosynthesis, *Phytochemistry*, 8, 697, 1969.
83. **Gauhl, E.,** Photosynthetic response to varying light intensity in ecotypes of *Solanum dulcamara* L. from shaded and exposed habitats, *Oecologia*, 22, 275, 1976.
84. **Patterson, D. T. and Duke, S. O.,** Effect of growth irradiance on the maximum photosynthetic capacity of water hyacinth *Eichhornia crassipes* (Mart.) Solms, *Plant Cell Physiol.*, 20, 177, 1979.
85. **Bunce, J. A., Patterson, D. T., Peet, M. M., and Alberte, R. S.,** Light acclimation during and after leaf expansion in soybean, *Plant Physiol.*, 60, 225, 1977.
86. **Blackman, G. E.,** Responses to environmental factors by plants in the vegetative phase, in *Growth in Living Systems*, Zarrow, M. X., Ed., Basic Books, New York, 1960, chap. 25.
87. **Patterson, D. T.,** Shading responses of purple and yellow nutsedges (*Cyperus rotundus* and *C. esculentus*), *Weed Sci.*, 30, 25, 1982.
88. **Patterson, D. T.,** Effects of shading and temperature on showy crotalaria *(Crotalaria spectabilis)*, *Weed Sci.*, 30, 692, 1982.
89. **Patterson, D. T.,** Shading effects on growth and partitioning of plant biomass in cogongrass (*Imperata cylindrica*) from shaded and exposed habitats, *Weed Sci.*, 28, 735, 1980.
90. **Patterson, D. T.,** The effects of shading on the growth and photosynthetic capacity of itchgrass *(Rottboellia exaltata)*, *Weed Sci.*, 27, 549, 1979.
91. **Mercado, B. L.,** Biology, problems, and control of *Rottboellia exaltata*, L. f., *Biotrop. Bull.*, 14, 1978.
92. **Thomas, P. E. L. and Allison, J. C. S.,** Competition between maize and *Rottboellia exaltata*, *J. Agric. Sci.*, 84, 305, 1975.
93. **Salisbury, F. B. and Ross, C. W.,** *Plant Physiology*, Wadsworth Publishing, Belmont, Calif., 1978, chap. 21.
94. **Kramer, P. J.,** *Water Relations of Plants*, Academic Press, New York, 1983.
95. **Pearcy, R. W.,** Temperature responses of growth and photosynthetic CO_2 exchange rates in coastal and desert races of *Atriplex lentiformis*, *Oecologia*, 26, 245, 1976.
96. **Billings, W. D., Godfrey, P. J., Chabot, B. F., and Bourque, D. P.,** Metabolic acclimation to temperature in arctic and alpine ecotypes of *Oxyria digyna*, *Arct. Alp. Res.*, 3, 277, 1971.
97. **Regehr, D. L. and Bazzaz, F. A.,** Low temperature photosynthesis in successional winter annuals, *Ecology*, 57, 1297, 1976.
98. **Williams, G. J., III,** Photosynthetic adaptation to temperature in C_3 and C_4 grasses, *Plant Physiol.*, 54, 709, 1974.
99. **Potter, J. R. and Jones, J. W.,** Leaf area partitioning as an important factor in growth, *Plant Physiol.*, 59, 10, 1977.
100. **Woledge, J. and Jewiss, O. R.,** The effect of temperature during growth on the subsequent rate of photosynthesis in leaves of tall fescue (*Festuca arundinaceae* Schreb.), *Ann. Bot. (London)*, 33, 897, 1969.
101. **Chandler, J. M.,** Competition of spurred anoda, velvetleaf, prickly sida, and Venice mallow in cotton, *Weed Sci.*, 25, 151, 1977.
102. **Oliver, L. R. and Lambert, B.,** Today's weed. Spurred anoda, *Weeds Today*, 5(3), 22, 1974.

103. **Patterson, D. T. and Flint, E. P.,** Effects of chilling on cotton (*Gossypium hirsutum*), velvetleaf (*Abutilon theophrasti*), and spurred anoda (*Anoda cristata*), *Weed Sci.*, 27, 473, 1979.
104. **Flint, E. P., Patterson, D. T., and Beyers, J. L.,** Interference and temperature effects on growth of cotton (*Gossypium hirsutum*), spurred anoda (*Anoda cristata*) and velvetleaf (*Abutilon theophrasti*), *Weed Sci.*, 31, 892, 1983.
105. **Pearcy, R. W., Tumosa, N., and Williams, K.,** Relationships between growth, photosynthesis and competitive interactions for a C_3 and a C_4 plant, *Oecologia*, 48, 371, 1981.
106. **Christie, E. K. and Detling, J. K.,** Analysis of interference between C_3 and C_4 grasses in relation to temperature and soil nitrogen supply, *Ecology*, 63, 1277, 1982.
107. **Flint, E. P. and Patterson, D. T.,** Interference and temperature effects on growth in soybean (*Glycine max*) and associated C_3 and C_4 weeds, *Weed Sci.*, 31, 193, 1983.
108. **Black, J. N.,** Competition between plants of different initial seed sizes in swards of subterranean clover (*Trifolium subterraneum* L.) with particular reference to leaf area and the light microclimate, *Aust. J. Agric. Res.*, 9, 299, 1958.
109. **Patterson, D. T., Meyer, C. R., Flint, E. P., and Quimby, P. C., Jr.,** Temperature responses and potential distribution of itchgrass (*Rottboellia exaltata*) in the United States, *Weed Sci.*, 27, 77, 1979.
110. **Patterson, D. T. and Flint, E. P.,** Effects of simulated field temperatures and chilling on itchgrass (*Rottboellia exaltata*), corn, (*Zea mays*), and soybean (*Glycine max*), *Weed Sci.*, 27, 645, 1979.
111. **Turner, N. C. and Kramer, P. J., Eds.,** *Adaptation of Plants to Water and High Temperature Stress*, John Wiley & Sons, New York, 1980.
112. **Hsiao, T. C.,** Plant responses to water stress, *Annu. Rev. Plant Physiol.*, 24, 519, 1973.
113. **Begg, J. E. and Turner, N. C.,** Crop water deficits, *Adv. Agron.*, 28, 161, 1976.
114. **Bourque, D. P. and Naylor, A. W.,** Large effects of small water deficits on chlorophyll accumulation and ribonucleic acid synthesis in etiolated leaves of jack bean (*Canavalia ensiformis* (L.) DC.), *Plant Physiol.*, 47, 591, 1971.
115. **Boyer, J. S.,** Water deficits and photosynthesis, in *Water Deficits and Plant Growth*, Vol. 4, Kozlowski, T. T., Ed., Academic Press, New York, 1976, chap. 4.
116. **Bradford, K. J. and Hsiao, T. C.,** Physiological responses to moderate water stress, in *Encyclopedia of Plant Physiology*, Vol. 12B, Pirson, A. and Zimmerman, M. H., Eds., Springer-Verlag, Berlin, 1982, chap. 9.
117. **Lawlor, D. W.,** Effects of water and heat stress on carbon metabolism of plants with C_3 and C_4 photosynthesis, in *Stress Physiology in Crop Plants*, Mussell, H. and Staples, R. C., Eds., John Wiley & Sons, New York, 1979, 303.
118. **Boyer, J. S. and McPherson, H. G.,** Physiology of water deficits in cereal crops, *Adv. Agron.*, 27, 1, 1975.
119. **Wiese, A. F. and van Diver, C. W.,** Soil moisture effects on competitive ability of weeds, *Weed Sci.*, 18, 518, 1970.
120. **Pavlychenko, T. K.,** Quantitative study of the entire root system of weed and crop plants under field conditions, *Ecology*, 18, 62, 1937.
121. **Weaver, J. E.,** *Root Development of Field Crops*, McGraw-Hill, New York, 1926.
122. **Davis, R. G., Johnson, W. C., and Wood, F. O.,** Weed root profiles, *Agron. J.*, 59, 555, 1967.
123. **Briggs, L. J. and Shantz, H. L.,** The Water Requirement of Plants. I. Investigations in the Great Plains in 1910 and 1911, Bur. Plant Ind. Bull. No. 284, U.S. Department of Agriculture, Washington, D.C., 1913.
124. **Briggs, L. J. and Shantz, H. L.,** The Water Requirement of Plants. II. A Review of the Literature, Bur. Plant Ind. Bull. No. 285, U.S. Department of Agriculture, Washington, D.C., 1913.
125. **Patterson, D. T. and Flint, E. P.,** Comparative water relations, photosynthesis, and growth of soybean (*Glycine max*) and seven associated weeds, *Weed Sci.*, 31, 318, 1983.
126. **Bunce, J. A., Miller, L. N., and Chabot, B. F.,** Competitive exploitation of soil water by five eastern North American tree species, *Bot. Gaz.*, 138, 168, 1977.
127. **Williams, E. D.,** Components of the vegetation of permanent grassland in relation to fertilizers and lime, *Ann. Appl. Biol.*, 83, 342, 1976.
128. **Banks, P. A., Santelmann, P. W., and Tucker, B. B.,** Influence of long-term soil fertility treatments on weed species in winter wheat, *Agron. J.*, 68, 825, 1976.
129. **Blackman, G. E. and Templeman, W. G.,** The nature of the competition between cereal crops and annual weeds, *J. Agric. Sci.*, 28, 247, 1938.
130. **Okafor, L. I. and DeDatta, S. K.,** Competition between upland rice and purple nutsedge for nitrogen, moisture, and light, *Weed Sci.*, 24, 43, 1976.
131. **Robinson, E. L.,** Yield and height of cotton as affected by weed density and nitrogen level, *Weed Sci.*, 24, 40, 1976.
132. **Volz, M. G.,** Infestations of yellow nutsedge in cropped soil: effects on soil nitrogen availability to the crop and on associated N transforming bacterial populations, *Agro-Ecosyst.*, 3, 313, 1977.

133. **Oliver, L. R. and Schreiber, M. M.,** Competition for CO_2 in a heteroculture, *Weed Sci.*, 22, 125, 1974.
134. **Keeling, C. D.,** The ocean and terrestrial biosphere as future sinks for fossil fuel CO_2, in *Interpretation of Climate and Photochemical Models, Ozone and Temperature Measurements*, AIP Conf. Proc. No. 82, Reck, R. A. and Hummel, J. R., Eds., American Institute of Physics, New York, 1982, 47.
135. **Baes, C. J., Goeller, H. E., Olson, J. S., and Rotty, R. M.,** *The Global Carbon Dioxide Problem*, ORNL-5194, Oak Ridge National Laboratory, Oak Ridge, Tenn., 1976.
136. **Wittwer, S. H. and Robb, W.,** Carbon dioxide enrichment of greenhouse atmospheres for crop production, *Econ. Bot.*, 18, 34, 1964.
137. **Akita, S. and Moss, D. N.,** Differential stomatal response between C_3 and C_4 species to atmospheric CO_2 concentration and light, *Crop Sci.*, 12, 789, 1972.
138. **Patterson, D. T. and Flint, E. P.,** Potential effects of global atmospheric CO_2 enrichment on the growth and competitiveness of C_3 and C_4 weed and crop plants, *Weed Sci.*, 28, 71, 1980.
139. **Patterson, D. T., Flint, E. P., and Beyers, J. L.,** Effects of CO_2 enrichment on competition between a C_4 weed and a C_3 crop, *Weed Sci.*, 32, 101, 1984.
140. **Kramer, P. J.,** Carbon dioxide concentration, photosynthesis, and dry matter production, *BioScience*, 32, 101, 1984.
141. **Sionit, N., Strain, B. R., and Hellmers, H.,** Growth response of wheat to CO_2 enrichment with different levels of mineral nutrition, *Agron. J.*, 73, 1023, 1981.
142. **Patterson, D. T. and Flint, E. P.,** Interacting effects of CO_2 and nutrient concentration, *Weed Sci.*, 30, 389, 1982.

Chapter 5

WEED ALLELOPATHY

Alan R. Putnam

TABLE OF CONTENTS

I. INTRODUCTION

Interference is the adverse effect that neighboring higher plants can exert on each other's growth.[1] Potential causes of this interference include: (1) *competition or allelospoly*:[2] the depletion of one or more limiting resources such as light, a nutrient, or water; (2) *allelopathy*:[3] the production of chemicals by living or decaying plant tissue which interferes with the growth of a neighboring plant; and (3) *indirect sources*, e.g., *allelomediation*:[4] includes effects on the physical or biological environment that interfere with the growth of a neighboring plant, e.g., harboring of herbivores that selectively graze certain species in the association.

Few, if any, investigations have definitively separated components of interference because of the complexity of this ecological phenomenon. Harper[5] indicated, "There is a need for a more rigorous philosophy in the search for cause and effect, perhaps like that adopted by pathologists and defined in Koch's postulates." Although field studies have often strongly implicated allelopathy, it has been virtually impossible to isolate the mechanisms in the field. In contrast, several laboratory studies have provided convincing evidence for allelopathy.[6-8]

The term *allelopathy* was coined by Molisch in 1937.[3] In current use, the term generally refers to the detrimental effects of higher plants of one species (the donor) on the germination, growth, or development of plants of another (recipient) species. Allelopathy can be separated from other mechanisms of plant interference in that the detrimental effect is exerted through release of a chemical by the donor species. Some investigators include stimulatory responses under the realm of allelopathy.[9] This is analogous to the special cases where herbicides can enhance growth at low concentrations, and in those instances, they are still classed as herbicides.

The fact that allelopathy can impact agriculture was apparently recognized by Democritus and Theophrastus in the fifth and third century B.C., respectively,[10] by deCandolle in 1832,[11] and more recently by many ecologists and agronomists.[9,12-15] In general, allelopathy has been related to problems with weed/crop interference,[16-19] with phytotoxicity in stubble mulch farming,[12,20] with certain types of crop rotations,[13,21,22] and with orchard replanting[23,24] or forest regeneration.[25,26] In many of these allelopathic interactions, it is not clear whether reduced crop production is due to direct effects of released toxins or whether the toxins may precondition the crop plant to invasions by plant pathogens. Rice[9] indicated that allelopathy may contribute to the weed seed longevity problem through: (1) chemical inhibitors in the seed preventing seed decay induced by microbes; (2) inhibitors functioning to keep seed dormant, but viable for many years.

Table 1
TERMS PREVIOUSLY USED TO DESCRIBE INHIBITORS IMPLICATED IN ALLELOPATHIC INVESTIGATIONS[34]

Term	Donor (agent)	Recipient (suscept)	Originator of term
Koline	Higher plant	Higher plant	Grummer
Marasmin	Higher plant	Microorganism	Gaumann
Phytoncide	Microorganism	Higher plant	Waksman
Antibiotic	Microorganism	Microorganism	Grummer

From Grummer, G., *Die gegenseitige Beeinflussung hoherer Pflanzen-Allelopathie,* G. Fischer, Jena, 1955. With permission.

There is extensive evidence that allelopathy may contribute to patterning of vegetation in natural ecosystems.[27-29] Distinct zones of inhibition are present under and adjacent to a variety of woody species, and often toxins from their litter are implicated.[30,31] One might speculate that aggressive perennial weed species that quickly gain dominance or form monocultures could exploit allelopathic mechanisms.

There are a variety of terms that have been assigned to inhibitors of biotic origin (Table 1). These have caused considerable confusion, because of their nonspecific or nondescriptive nature. Our laboratory has recently proposed that the term (1) *phytoinhibitins* be used to describe compounds of higher plant origin that inhibit other plants; and (2) *saproinhibitins* be used to describe compounds of microbial origin that are toxic to higher plants.[32] These terms are more descriptive than the terms koline and phytoncide.

Allelopathy is a substituent of a higher-level order of chemical ecology. Whittaker and Feeny[33] have classed interspecies allelochemic effects on the basis of whether the adaptive advantage goes to the donor or recipient. Allomones, which give adaptive advantage to the producer, include repellants, escape substances, suppressants, venoms, inductants, counteractants, and attractants. Allelopathic chemicals could be classified as suppressants. Some inhibitors of plant origin may also display intraspecific effects (autotoxicity).

II. PROOF OF ALLELOPATHY

Although there are numerous studies where excellent evidence for allelopathy is presented, seldom has an investigation followed a set of criteria similar to Koch's postulates to provide undisputable proof.[32] To achieve convincing proof, the following steps are, perhaps, necessary:

1. Demonstrate the existence of interference, describe the symptoms, and quantitate the degree of interference.
2. Isolate, assay, characterize, and synthesize the toxins. Isolation of toxic substances is a key step in obtaining direct evidence that the interference is caused by chemicals.
3. The symptoms of interference previously diagnosed should be repeated by application of the toxin(s) at rates present in nature at the appropriate stages of suscept growth. This will show that symptoms can, indeed, be reproduced (quantitatively, if possible) by the toxin or combination of toxins.
4. The release, movement, and uptake of the toxin should be monitored and shown to be sufficient for the observed interference. Meeting this criterion would allow the conclusion that the toxin is released in sufficient concentration to move across the environmental barriers, whether through soil or air, and still be taken up in quantities sufficient to cause toxicity in the suscept.

It would be desirable but not essential to consider the selectivity of the toxin by comparing the range of species affected by the isolated toxin to the range of species that show allelopathic injury when grown with the agent. In such a study of selectivity, the autotoxicity of the toxin should be studied as well. Such a selectivity study has been conducted by Stowe[35] by correlating susceptibility to crude toxin preparations with spatial relationships of species in the field.

In nature, several compounds may act jointly to exert this allelochemical effect. Synergistic effects have already been implicated for organic acids and phenolic compounds.[36,37]

III. SOURCES OF ALLELOPATHIC CHEMICALS

Chemicals with allelopathic potential are present in virtually all plant tissues, including leaves, stems, roots, rhizomes, flowers, fruits, and seeds. Whether these compounds are released by the plant in quantities sufficient to elicit a response remains the critical question in studies of allelopathy. Allelochemics may be released from plant tissues in a variety of ways, including volatilization, root exudation, leaching, and decomposition of the plant residues.

A. Volatilization

Reports of the release of volatile toxins originate primarily from studies on plants found in more arid regions of the world. Among the genera shown to release volatiles are *Artemisia*,[8] *Eucalyptus*,[31] and *Salvia*.[38] When identified, the compounds were identified as terpenoid compounds, mainly monoterpenes and sesquiterpenes. Work of Muller and associates[38] has indicated that these compounds may be absorbed in vapor by surrounding plants, be absorbed from condensate in dew, or they may reach the soil and be taken up by plant roots.

B. Root Exudation

A myriad of compounds are released by plant roots.[39,40] Whether the compounds are actively exuded, leaked, or arise from dead cells sloughing off the roots is not clearly understood at this time. Much of the evidence for root-mediated allelopathy has come from studies where solutions passing by root systems of one plant are recycled into media containing the indicator species. Often, these experiments are set up in staircase fashion so that the solutions may move by gravity through a series of interconnected pots.[16] Common lambsquarters (*Chenopodium album* L.) plants were found to exude toxic levels of oxalic acid into solution when they reached the flowering stages.[41] A more recent study by Tang and Young[6] successfully utilized an adsorptive column (XAD-4) to selectively trap organic and hydrophobic root exudates while allowing nutrient ions and other hydrophilic compounds to pass through (Figure 1). They were able to identify 16 compounds that were exuded from the roots of Bigalta limpograss (*Hemarthia altissima*). These compounds represented a variety of benzoic, cinnamic, and phenolic acids. There are only a few instances where either radiolabeled precursers of allelochemics, allelochemics themselves, or labeled growth-regulating compounds have been applied to plants after which the label is monitored in the nutrient solution or recipient plant.[40,42,43] This technique should prove helpful to quantify root exudation of compounds in future studies.

C. Leaching

A variety of chemicals may be leached from the aerial portions of plants by rainwater or by fog-drip.[44] Among compounds shown to be leached from plants are organic acids, sugars, amino acids, pectic substances, gibberellic acids, terpenoids, alkaloids, and phenolic compounds. Leachates of *Chrysanthemum* leaves were shown to be toxic,[44] and foliar leachates may play an important role in failure of understory plants to grow in forest stands.[30,31] It

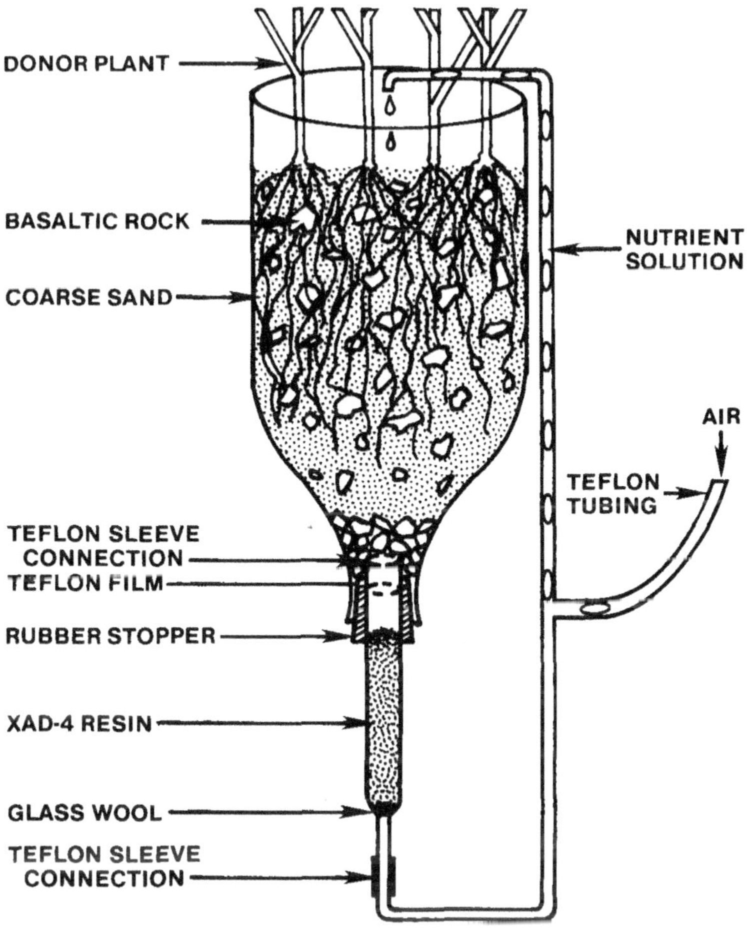

FIGURE 1. This trapping device developed by Tang and Young[6] allows collection of substances exuded by undisturbed plant roots. The continuously circulated solution, which contains mineral nutrients required for growth, elutes allelopathic substances from the media, then passes through an XAD-4 adsorptive resin, which traps the allelochemicals. The resin column then can be eluted to remove the chemicals for bioassay and characterization. (From Tang, C. S. and Young, C. C., *Plant Physiol.*, 69, 155, 1982. With permission.)

has often been observed that yields of flax (*Linum usitatissimum* L.) are greatly reduced when a few plants of flax weed (*Camelina alyssum*) are present in the field. Grummer and Beyer[45] leached inhibitors from this weed that could inhibit flax growth by 40%. Recently, Colton and Einhellig[17] suggested that leaf leachates of velvetleaf (*Abutilon theophrasti* Medic.) may be inhibitory to soybean (*Glycine max* [L.] Merr.).

D. Decomposition of Plant Residues

After death of the plant or its selected organs, chemicals may be released directly by leaching of the litter. Loss of membrane integrity that previously imposed permeability barriers now may allow release of a variety of compounds that may impose their toxicity in an additive or synergistic matter. In addition to direct release of compounds from the tissue, microbes in the rhizosphere may induce production of toxic compounds by enzymatic degradation of conjugates or polymers present in the plant tissue. Examples of this phenomenon are the action by microbes on the cyanogenic glycosides of johnsongrass (*Sorghum halepense* (L.) Pers), and Prunus species to produce two toxins, HCN, and the corresponding benzaldehydes.[46,47]

(a) *Sorghum species*

$$\text{dhurrin} \xrightarrow{\text{emulsin}} \text{HCN} + p\text{-hydroxybenzaldehyde}$$

(b) *Prunus species*

$$\text{amydalin} \xrightarrow{\text{emulsin}} \text{HCN} + \text{benzaldehyde}$$

The toxicity arising from plant residues undoubtedly provides some of the more challenging problems and opportunities for agronomists and weed scientists. Stubble-mulch farming has been practiced for decades in the plains states for soil and water conservation. Toxins from the stubble have provided toxicity to certain rotational crops.[13,20] There is currently a pronounced trend toward more conservation tillage (including no-tillage) practices, which preserve surface plant residues. Not only can the residues have an influence on crop emergence, growth, and productivity, but they can also influence similar aspects of weed growth. Our recent work indicates that management of selected crop residues can greatly reduce weed germination and growth.[48,49]

IV. WEED SPECIES WITH ALLELOPATHIC POTENTIAL

The number of weed species already alleged to have allelopathic potential is relatively large (Table 2). Allelopathy could contribute toward the aggressiveness of weeds in weed: crop interference or in patterning of weeds in a field through influences of their exudates, leachates, volatiles, or litter products. Although weed scientists have performed numerous weed/crop "competition" studies, seldom has allelopathy been considered, mentioned, or eliminated as a possible mechanism. Even worse, many investigators have failed to recognize the inherent differences in these mechanisms. Proof of competitive interference is as necessary as proof of allelopathy.[32] Seldom has an investigator proven which, if any, resource(s) becomes limiting in these "competition" experiments.

There is considerable evidence now accumulated to suggest that some of the more aggressive perennial weed species, including quackgrass (*Agropyron repens* (L.) Beauv.,[50] Canada thistle (*Cirsium arvense*) (L.) Scop.,[51] johnsongrass,[52] and yellow nutsedge (*Cyperus esculentus* L.),[53] may impose allelopathic influences, particularly through toxins released from their residues. There are also several annual weed species in which allelopathy is implicated. Perhaps best documented is giant foxtail (*Setaria faberi* Herrm.), whose residues have been shown to severely inhibit the growth of corn (*Zea mays* L.).[16,54] More chemical isolation and identification work is necessary to confirm most of these observations. For example, many of the reports contained in Table 2 are based only on activities of plant extracts or residues. Extracts of almost any plant will be toxic at some concentration, therefore, more detailed work is needed on a number of these species.

In addition to the above weed species, allelopathy has been documented with a number of important crops. Certain crop residues may exert toxicity on succeeding crops. In addition, there have been attempts to find crop types which themselves have superior interference ability with weeds. This information will be considered in more detail later in the chapter.

V. NATURAL PRODUCTS IDENTIFIED AS ALLELOPATHIC AGENTS

The inhibitors from plants and their associated microbes represent a myriad of chemical compounds from the extremely simple gases and aliphatic compounds to complex multiringed

Table 2
COMMON WEED INHABITANTS WITH ALLEGED ALLELOPATHIC ACTIVITY IN AGROECOSYSTEMS

Scientific name	Common name	Susceptible species	Ref.
Abutilon theophrasti Medic.	Velvetleaf	Several	17,55
Agropyron repens (L.) Beauv.	Quackgrass	Several	18,50,56
Agrostemma githago L.	Corn cockle	Wheat[a]	57
Allium vineale L.	Wild garlic	Oats	58
Amaranthus spinosus L.	Spiny amaranth	Coffee	59
Ambrosia artemisiifolia L.	Common ragweed	Several	60
A. psilostachya D.C.	Western ragweed	Several	71
A. trifida L.	Giant ragweed	Pea, wheat	62
Artemisia absinthium L.	Absinth wormwood	Several	27,63
A. vulgaris L.	Mugwort	Cucumber	79
Asclepias syriaca L.	Common milkweed	Sorghum	19
Avena fatua L.	Wild oat	Several	29
Brassica sp.	Mustard	Several	1
Camelina alyssum	Flax weed	Flax	45
Celosia argentea		Bajra	65
Chenopodium album L.	Common lambsquarters	Cucumber, oats, maize	41
Cirsium arvense (L.) Scop.	Canada thistle	Several	51,66
Cynodon dactylon (L.) Pers.	Bermudagrass	Coffee	59
Cyperus esculentus L.	Yellow nutsedge	Maize	53,67
C. rotundus L.	Purple nutsedge	Sorghum, soybean	68,69
Digera arvensis		Bajra	70
Digitaria sanguinalis (L.) Scop.	Large crabgrass	Pioneer weeds	72
Erica scoparia	Heath	Red clover	73
Euphorbia esula L.	Leafy spurge	Pea, wheat	61,74
E. supina Raf.	Prostrate spurge	Several	75
Helianthus annuus L.	Sunflower	Several	9,140
H. mollis		Radish, wheat	76
Hemarthia altissima	Bigalta limpograss	Desmodium	6
Holcus mollis L.	Velvetgrass	Barley	77
Imperata cylndrica (L.) Beauv.	Alang-alang	Several	64
Lolium multiflorum Lam.	Italian ryegrass	Oats, brome, lettuce, clover	78
Matricaria inodora	Mayweed	Barley	79
Nepeta cataria L.	Catnip	Peas, wheat	62
Parthenium hysterophorus	Ragweed parthenium	Several	80,81
Poa spp.	Bluegrass	Tomato	82
Polygonum orientale L.	Princesfeather	Mustard	83
P. persicaria L.	Ladysthumb	Potato, flax	84
Portulaca oleracea L.	Common purslane	Peas, wheat	62
Rumex crispus L.	Dock	Corn, pigweed, sorghum	85
Saccharum spontaneum	Wild cane	Wheat	86
Salsola kali L.	Russian thistle	Several	87
Schinus molle L.	California peppertree	Cucumber, wheat	88
Setaria faberi Herrm.	Giant foxtail	Maize	16,54
Solidago sp.	Goldenrod	Several	62
Sorghum halepense (L.) Pers.	Johnsongrass	Pioneer weeds	46
Stellaria media (L.) Cyrill.	Common chickweed	Barley	89

[a] Stimulated growth.

Reprinted with permission from Putnam, A. R., *Chem. Eng. News*, 61(19), 38, 1983. Copyright American Chemical Society.

Chemical name	Class	Structural formula	Natural source
Dhurrin	Cyanogenic glucoside	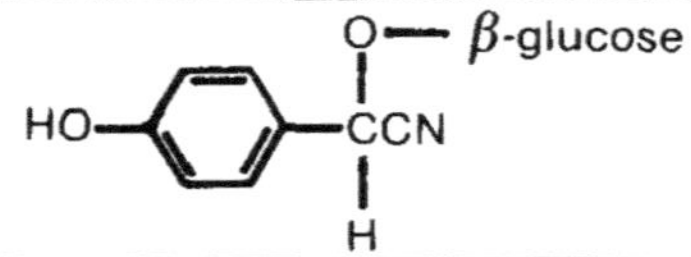	Sorghum plants
Allylisothiocyanate	Thiocyanate	$CH_2 = CHCH_2NCS$	Mustard plants
Camphor	Monoterpene	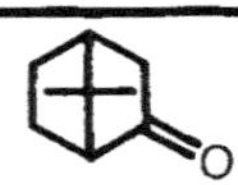	*Salvia* Shrubs
Acetic acid	Aliphatic acid	CH_3COOH	Decomposing straw
Cinnamic acid	Aromatic acid	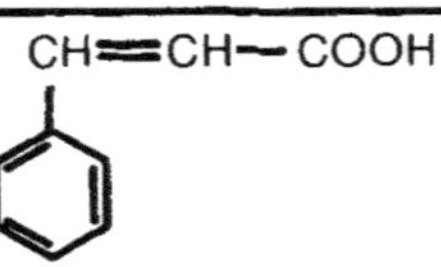	Guayule plants
Arbutin	Phenolic compound	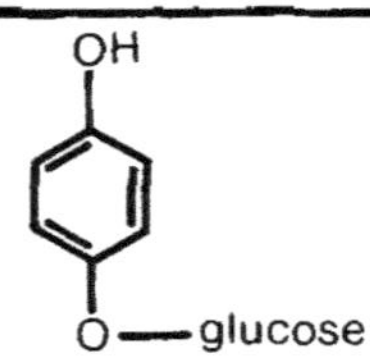	Manzanita shrubs
Bialaphos	Amino acid derivative	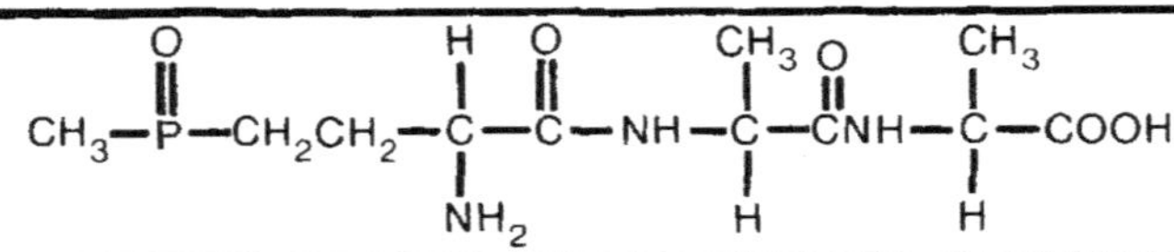	*Streptomyces hygroscopicus*
Patulin	Simple lactone	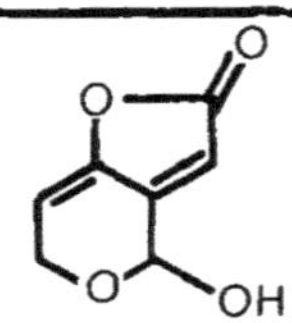	*Penicillium* fungus on wheat straw

Psoralen	Furanocoumarin	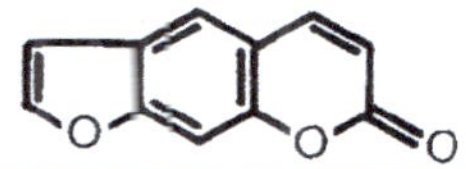	*Psoralea* plants
Juglone	Quinone	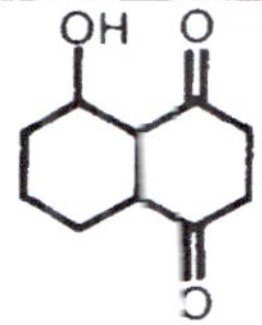	Black walnut trees
Phlorizin	Flavonoid	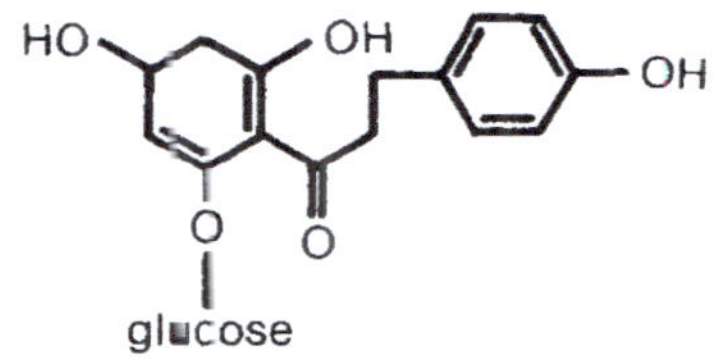	Apple roots
Gallic acid	Tannin	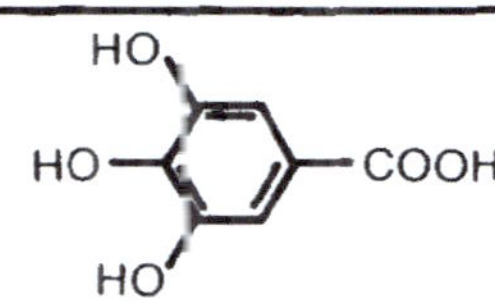	Spurge plants
Caffeine	Alkaloid	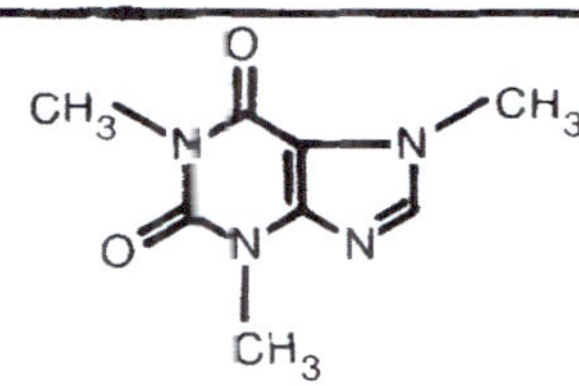	Coffee plants

FIGURE 2. Examples of diverse chemistries implicated in allelopathic interactions. Both higher plants and microorganisms are sources of these chemicals.

aromatic compounds (Figure 2). There are only a few weed species in which the nature of the secondary metabolites have been characterized. For this reason, it may be more useful to consider allelopathic chemicals from species studied in more detail.

The groups of compounds implicated in allelopathy have been divided into chemical classes by recent reviewers.[9,90,91] I have arbitrarily classed these chemicals into (A) toxic gases, (B) organic acids and aldehydes, (C) aromatic acids, (D) simple unsaturated lactones, (E) coumarins, (F) quinones, (G) flavonoids, (H) tannins, (I) alkaloids, (J) terpenoids and steroids, and (K) miscellaneous and unknowns. Although many of these compounds are secondary products of plant metabolism, several are degradation products that occur in the presence of microbial enzymes.

A. Toxic Gases

Tissues of many plants contain relatively high concentrations of cyanogenic glycosides, e.g., amygdalin, dhurrin, and linamarin.[92] Upon hydrolysis, HCN may be liberated from these compounds, a reaction known as cyanogenesis. Cyanide has been demonstrated to prevent germination of a variety of seeds[93] and may also inhibit root growth when released from adjacent decaying roots.[24] Realizing the toxicity of cyanide groups may have provided a basis for evaluation of various synthetic azides and thiocyanates as herbicidal compounds.

Several researchers have also demonstrated that NH_3 gas inhibits germination and is toxic to seedling growth. For example, toxic levels of ammonia may accumulate upon hydrolysis reactions during germination of sugarbeet (*Beta vulgaris* L.) and of lettuce (*Lactuca sativa* L.) seed.[94] Ammonia is thought to be even more inhibitory to germination than cyanide.[93] Ethylene gas is liberated by many plant tissues and is produced in ripening fruits in relatively high quantities. For example, apples may produce up to 11 mℓ/kg/24 hr.[95] This compound has pronounced effects both on seed germination and plant growth.[36] Its stimulatory action on seed germination at low concentrations has been exploited for control of witchweed (*Striga lutea* Lour.), a serious parasitic weed of the southeastern U.S.[96]

Many plants belonging to the mustard (Cruciferae) family contain strong volatile inhibitors of germination and growth.[9] The so-called mustard oils, allyl isothiocyanate and β-phenethyl-isothiocyanate, common throughout tissues of *Brassica* and *Sinapsis*, are potent inhibitors of germination as well as inhibitors of microbial activity.[36]

Several plant species that inhabit desert or chaparral ecosystems are known to release volatile chemical inhibitors. Some classical work in this area has been performed by Muller and associates at the University of California at Santa Barbara. They isolated several volatile monoterpenes including camphor and cineole from *Salvia leucophylla* Greene, and later demonstrated numerous herbicidal effects of these compounds.[38] They suggest these volatiles can contribute to inhibition of annual plants around *Salvia* shrubs. These and related volatile compounds are also released from *Artemisia californica* Less. and may provide the 1- to 2-m zone of inhibition around these shrubs.[8]

B. Organic Acids and Aldehydes

Aliphatic acids (several from the Kreb's cycle) are well-known inhibitors of seed germination. Their toxic effects can be separated from that caused by lowered pH.[36] Malic and citric acid, which are common components of fruit juices and plant sap, inhibit germination of many seeds at concentrations of 0.1 to 1%. A variety of tricarboxylic acids are present in sorghum (*Sorghum bicolor* L.) herbage and contribute to toxicity of sorghum residues.[97] Since different acids almost always occur in mixtures in juices and sap, their synergistic effects cannot be ignored. For example, a mixture of 0.025% citric, 0.025% tartaric, and 0.01% salicylic acid (total acids 0.06%) exerts much more inhibition than any of the acids applied alone at 0.06%.[45]

Simple aliphatic acids are formed upon anaerobic decomposition of plant residues in soil.

For example, acids are released from decaying quackgrass residues and the straw of cereal grains.[98,99] Although acetic acid is one of the least toxic compounds, it may still be one of the most important, since it is produced in large quantities; and under low O_2 conditions, it may remain long enough to inhibit seed germination.[99] Acetaldehyde also causes inhibition of seed germination.[36]

Some of the first synthetic herbicides (some of which are still in use) were halogenated analogues of aliphatic acids. Additions of chlorine to these molecules lended more activity and additional soil persistence. Similarly, slight modifications in the structures of many natural products could provide characteristics that improve their utility of herbicides.

Although certain amino acids have also been implicated in allelopathy, there is no clear understanding of which, if any, are important in nature. Moje[100] has summarized the work of numerous investigators who have reported inhibitory effects of amino acids in laboratory experiments. Several amino acids were found in the toxic leachates from velvetleaf seeds.[55,101]

C. Aromatic Acids

A variety of aromatic acids, aldehydes, and their related phenols have also been implicated in allelopathy. In particular, cinnamic and benzoic acid derivatives have been isolated from a wide range of plant species, their residues, and surrounding soils.[9,91] Cinnamic acid derivatives commonly reported in allelopathy studies are chlorogenic, *p*-coumaric, ferulic, and caffeic acids. These compounds may be derived from the aromatic amino acids phenylalanine or tyrosine through the shikimic acid pathway. *p*-Hydroxybenzoic, syringic, and vanillic acids are some commonly reported benzoic acids from a variety of plants and soils. These compounds have been implicated as soil toxins where residues of plants (particularly wheat, corn, sorghum, and oats) remain on the soil.[12,20]

Phytotoxic aromatic aldehydes are produced upon the breakdown of cyanogenic glucosides. For example, *p*-hydroxybenzaldehyde is a toxic product upon breakdown of dhurrin in various *Sorghum* species. Similarly, benzaldehyde is produced upon breakdown of amygdalin in peach (*Prunus persica* [L.] Batsch.) roots. Benzaldehyde and HCN may both enter the vapor phase and act together as toxins against a variety of plants.

D. Simple Unsaturated Lactones

Several simple lactones (those arising from acetate) are potent inhibitors of seed germination. For example, parasorbic acid from the fruits of Mt. Ash (*Sorbus aucuparia* L.) is inhibitory to *Lepidium* seed germination and also demonstrates antibacterial activity.[36] Patulin, an extremely phytotoxic microbial product, is produced by a number of soil fungi including *Penicillium urticae*.[102] This organism produces large quantities of patulin when growing on residues of wheat (*Triticum aestivum* L.). Patulin is extremely toxic to both plants and animals. Its mammalian toxicity would probably exclude its acceptability as a marketable herbicide, but, nevertheless, it does occur naturally and causes phytotoxicity in agroecosystems.

E. Coumarins

Coumarins are lactones of *o*-hydroxycinnamic acid whose side chains are often isoprenoids. Coumarin, esculin and psoralen (a furocoumarin) are all potent inhibitors of seed germination.[9] Inhibitors of this group are commonly produced by legumes and cereal grains. For example, coumarin is produced by sweet clover (*Melitotus alba* Desr.),[103] esculin is produced by timothy (*Phleum pratense* L.) roots,[104] and scopoletin or its 7-glucoside, scopolin, is produced by a variety of plants.[9] In a screening study to detect allelopathic oat (*Avena sativa* L.) lines, this characteristic was attributed to their ability to exude fluorescent scopoletin plus related compounds.[105] Psoralen, a furocoumarin that has been isolated from seeds of *Psoralea subacaulis* and fruits of several species, has been shown to inhibit germination of lettuce seeds at concentrations as low as 0.001 ppm. This compound also inhibits microbial growth. The extremely high activity in members of this chemical group indicate that they may be excellent sources of herbicidal chemistries.

F. Quinones

Only one quinone of plant origin has demonstrated toxicity on other plants. Juglone, a potent inhibitor from black walnut (*Juglans nigra* L.), was identified as 5-hydroxynaphthoquinone in 1928.[106] This compound is extremely toxic to many herbaceous crop plants including tomato (*Lycopersicon esculentum* Mill.) and alfalfa (*Medicago sativa* L.) and adjoining tree crops, including apple. This compound can apparently be released from leaves and fruits as well as the bark of roots.

Numerous other naphthoquinones, anthraquinones, and dimeric quinones are produced by higher plants and fungi. Few, if any, have been assayed for their herbicidal activity, although some of the fungal products may produce adverse effects on plants.[9] Phytopathology literature indicates that quinones, formed upon oxidation of phenolics by polyphenol oxidase, may contribute to resistance of plants to plant pathogens. This group of compounds deserves further investigation because synthetic chemicals that are quinones possess diverse pesticidal activities.

G. Flavonoids

The flavonoids are a diverse group of compounds that are widespread in higher plants. The insecticide rotenone is a naturally occurring isoflavonoid. Only a few have yet to be implicated in allelopathy, although many researchers believe that this chemical group will unveil a number of important compounds. This group poses some challenging analytical problems in that many are complex structures, several of which appear to be glycosides. One important compound in this group is phlorizin, the 6-glucoside of phloretin, an inhibitor isolated from apple (*Malus sylvestris* Mill.) roots. It is directly toxic to apple seedlings, and several of its breakdown products also demonstrate toxicity.[23] Rice and associates[9] have isolated a number of flavonoids from climax vegetation associations (tall prairie and oak forest) and found these compounds to be extremely inhibitory to both nitrifying bacteria and to seed germination. Among the inhibitors isolated from bedstraw (*Galium mollugo* L.) was the flavonoid, diosmetin trioside.[107] This compound may contribute to the allelopathic influences reported with that plant.

H. Tannins

This grouping includes both hydrolyzable and condensed tannins. Several of the more common hydrolyzable tannins are sugar esters of gallic acid, while others are complex mixtures of several phenolic acids. Hydrolyzable tannins have been implicated as seed germination inhibitors, inhibitors of nitrogen-fixing and nitrifying bacteria, and inhibitors of plant growth.[9] Many plant residues (particularly those from woody plants) contain gallic, ellagic, or digallic acids.[108] Much work is needed to determine the importance of these compounds and the lesser-understood condensed tannins in allelopathic interactions. Since many of these compounds inhibit nitrification, they could provide important products for agriculture. One synthetic product, nitrapyrin, is already marketed for this purpose.

I. Alkaloids

The alkaloids are cyclic compounds containing nitrogen in their rings or side chains. There has been little recent work on the importance of alkaloids in allelopathy, although Evanari pointed out back in 1949 that alkaloids are potent inhibitors of seed germination.[36] They are particularly important inhibitors of tobacco (*Nicotiana tabacum* L.), coffee (*Coffea arabica* L.), and cocoa (*Theobroma cacao* L.) seeds. Evanari indicated that the strongest alkaloid inhibitors were cocaine, physostigmin, caffeine, chinin, chinchonin, cinchonidin, tropa acid, strychnin, berberin, and codein. Selective herbicidal action of caffeine has recently been reported.[109] Several microbial alkaloids, including fusaric acid and α-picolinic acid, also have reported toxicity to plants.[110] One of the more active synthetic herbicides now on the market, picloram, is a chlorinated picolinic acid derivative.

J. Terpenoids and Steroids

These compounds are built from 5-carbon isoprene units linked together in a variety of configurations that differ in ring closure, functional group, and degree of saturation. The monoterpenoids (C_{10}) are the major "essential oils" of plants and are also the major group of inhibitors that have been identified in this class of compounds. Previously mentioned were the volatile inhibitors camphor and cineole, which, along with camphene, dipentene, α-pinene, and β-pinene, are the compounds produced by the various *Salvia* species.[38] These may also contribute to allelopathy in *Eucalyptus* and *Artemisia*.[8] Several fungi also produce terpenoids that are believed to cause the destruction of tissue or produce lesions on higher plants.[110]

The phytotoxic effects of steroids on higher plants are not well documented. However, two compounds in particular, digitoxigenin and strophanthidin, are known to have strong antimicrobial activity.[36]

K. Miscellaneous and Unknowns

There are a few additional compounds implicated in allelopathy including long-chain fatty acids, alcohols, polypeptides, and nucleosides.[9,91] Most allelopathic chemicals are believed to be secondary products that serve a number of protective roles for the plant.[33,111] Among these are various resistances to pest infection.

New chemicals are being isolated from plants and microorganisms daily. Swain[111] recently reported that over 10,000 low-molecular weight secondary products have been isolated from higher plants and fungi. In addition, he proposed that the total number might approximate 400,000 chemicals. These chemicals or their analogues could provide important new sources of herbicide chemistry in the future. There is considerable interest in the industry on at least two approaches involving allelochemics for weed control. One involves the production of crops (perhaps through genetic engineering), which can themselves suppress associates or provide a source of chemicals. Another approach is to produce the natural herbicides through batch culture with microorganisms.

VI. METHODS FOR ISOLATION, BIOASSAY, AND IDENTIFICATION

A. Extraction or Leaching from Tissue

The techniques utilized in characterization of natural products evolve rapidly as more sophisticated instrumentation is developed. There is probably no standard technique that will work effectively for every compound. Briefly, isolation of a compound involves extraction or collection in an appropriate solvent or adsorbant.[14] A commonly used extraction solvent is cold-water or aqueous methanol in which either dried or live plant parts are soaked. After extracting the material for varying lengths of time, the exuded material is usually filtered or centrifuged before bioassay. In other described cases of cold-water extraction, the plant parts are macerated and placed in dishes containing a moistened cellulose sponge that supports a filter paper on which indicator seeds are sown.[31] Thus, cold-water extraction and bioassay are carried out simultaneously. Users of this method imply that it simulates natural release of compounds by the action of rain on the standing or fallen plant material.

Other methods of extracting plant materials have involved boiling water, autoclaving, or organic solvent extraction.[14] Hot water or autoclaving is used to increase the diffusion of soluble chemicals into an aqueous phase and to eliminate microbial decay. The choice of extraction solvents depends on whether one is looking for compounds that would realistically be released in nature or whether one wishes to remove all toxic compounds. In any of these methods, the phytotoxic material isolated may not necessarily be the same one exerting its influence under natural conditions.

B. Root Exudates

Several assay techniques have been used to detect the presence of inhibitory substances in root exudates. These have included (1) using agar as a medium for plant growth and examining for the presence of "active" chemicals left in the agar;[112] (2) using sand in which both donor and recipient plants are present and evaluating the effects early in plant development before competition for growth factors occurs;[113] (3) growing donor plants in sand for specific times, leaching the sand, and evaluating the leachate on recipient plants in petri dishes, soil, or sand;[105,113] (4) growing donor and recipient plants in sand in pots arranged in the so-called staircase system in which the pots are alternated so that the nutrient solution flows from the donor to the recipient and back to a reservoir and back again for varying periods of time each day;[16,72] and (5) collecting exudates on adsorptive columns and eluting them for bioassay.[6]

C. Release from Plant Litter

In studies to examine the release of organic substances from decomposing matter, either live or dried plant material is placed in or on soil for a selected period of time before bioassaying with recipient plants. Some studies have involved isolating specific fungi involved in decomposing the plant material and evaluating by-products of fungal metabolism on plant growth.[102] As pointed out by Rice[9] in studies with plant litter in soil, it is difficult to determine whether the toxic agent comes from the plant, the microorganism, or is a result of an additive or synergistic effect of both. Short-term experiments with autoclaved vs. nonautoclaved media could be helpful to differentiate these effects.

D. Volatiles

Methodology for examining volatile allelopathic products has received considerable attention by Muller and associates.[8] Their method involves germination of seeds of indicator plants on or between filter paper sheets on a cellulose sponge that is placed in a larger container adjacent to beakers containing the donor plants. The only contact between plant material and seed is aerial. This technique has given results similar to those observed under natural conditions.

E. Separation and Characterization of Chemicals

Chemical separations may first be accomplished by partitioning on the basis of polarity into a series of solvents from nonpolar hexane to more polar compounds like *n*-butanol. Compounds may also be separated by molecular size, charge, or adsorptive characteristics, etc. Various chromatography methods are utilized, including columns, thin-layer (TLC) gas-liquid (GLC), and, more recently, high-pressure liquid (HPLC) systems. HPLC has proven particularly useful for separations of water-soluble compounds from relatively crude plant extracts. At one time, the major effort toward compound identification involved chemical tests to detect specific functional groups. Now characterization is usually accomplished using a series of spectroscopic analyses. Initially, ultraviolet spectroscopy (UV) is useful in this regard to detect specific functional groups. More recently, infrared (IR) spectroscopy and nuclear magnetic resonance (NMR) have helped immensely in determining natural products structure by indicating the functional groups and relative positions of atoms. Mass spectrometry is a relatively recent addition to the analytical arsenal that provides additional clues as to molecular size and composition. It can quickly provide confirmation of complex organic molecular structures. Tandem mass spectrometry (MS/MS) or GC-MS are recent developments which also allow analyses of mixtures of compounds. Effective studies of allelopathy should now involve natural products' chemists who can provide structure elucidation with the powerful instrumentation now available.

VII. FACTORS AFFECTING PRODUCTION OF ALLELOCHEMICS

Plants vary in their production of allelopathic chemicals depending upon the environment in which they are grown and the stresses that they encounter. One difficulty faced by researchers is that greenhouse-grown plants may produce only limited quantities of inhibitors. Ultraviolet (UV) light is absent in closed greenhouses, and several investigations have shown that UV light greatly enhances the production of allelopathic chemicals.[9] For example, when greenhouse light was supplemented by UV, the chlorogenic acid content of tobacco was increased sixfold, and approximated that of plants grown outdoors.[114] Sunflower *Helianthus annuas* L.) plants receiving supplemental UV radiation also produced more scopolin and chlorogenic acids.[115]

Light quality, intensity, and duration all seem to be important factors regulating the synthesis of allelopathic chemicals. For example, tobacco plants exposed to red light at the end of the day produced more alkaloids but fewer phenolics than those exposed to far-red.[116] Long days seem to enhance the content of phenolic acids and terpenes in a variety of plants.[9]

Nutrient deficiencies may also influence the production of allelochemics. The compounds studied in great detail have been the phenolic compounds and scopolin-related chemicals. Deficiencies of boron, calcium, magnesium, nitrogen, phosphorus, potassium, and sulfur have all been reported to enhance the concentration of chlorogenic acids and scopolin in a variety of plants.[9] In some species, chlorogenic acids have decreased in plants that are deficient in magnesium or potassium.

Water stress alone or in combination with other stresses can induce substantial increases in concentrations of chlorogenic and isochlorogenic acids in plants. For example, in sunflower, the combination of drought stress and nitrogen deficiency provided approximately 15-fold increases in these compounds.[117] Heat and chilling have also been shown to influence the production and release of allelochemics.[9]

Water stress alone or in combination with other stresses can induce substantial increases in concentration of chlorogenic and isochlorogenic acids in plants. For example, in sunflower, the combination of drought stress and nitrogen deficiency provided approximately 15-fold increases in these compounds.[117] Heat and chilling have also been shown to influence the production and release of allelochemics.[9]

There are numerous other factors that may influence the synthesis or release of chemicals by plants. Stress by exposure to herbicides or other natural product inhibitors may induce production of scopolin-related compounds.[118] The type and age of plant tissues are extremely important since compounds are not uniformly distributed in the plant. Between species, there are great differences in ability to produce allelochemics. Within species, differences may exist in the amount of toxin produced by different genotypes. For example, various oat lines show differences in their ability to exude scopoletin and related compounds.[105] Some cucumber (*Cucumis sativus* L.) accessions were found to greatly inhibit weed germination while others had no effect, or even stimulated growth.[113] The implications of all these findings are that we may either select plant types that are more allelopathic, or perhaps enhance inhibitor production by exerting the proper stresses on the plants.

VIII. MODE OF ACTION OF INHIBITORS

This area of research has caused similar challenges for investigators working with either natural products or synthetic herbicides. The major difficulty is to separate secondary effects from primary causes. Although effects can be measured in isolated systems, there always remains the critical questions of whether the inhibitor reaches that site in the plant in sufficient concentration to specifically influence that reaction, and whether other processes may be affected more quickly.

A. Effects on Nutrient Uptake

There is considerable evidence that allelopathic chemicals can alter the rate of ion uptake by plants. In fact, this may be the underlying mechanism behind many plant interference responses that have been labeled "competition". For example, salicylic acid, a phenolic compound, has been shown to inhibit the uptake of potassium by plants. Both salicylic acid and ferulic acid were both inhibitory to K^+ uptake by oat roots, particularly at lower pHs.[119] Other recent work indicated that salicylic acid uptake by plants is also greater at low pH.[120] Numerous other studies with whole plants and cell cultures have indicated a reduction in uptake of both macro- and micronutrients in the presence of phenolic acids.[9]

Chambers and Holm[121] measured uptake of ^{32}P by beans (*Phaseolus vulgaris* L.) growing with other beans or redroot pigweed (*Amaranthus retroflexus* L.) and found that although pigweed used seven times as much P, it had less effect on P uptake by beans than other bean plants. Buchholtz[122] observed that the presence of quackgrass produced N- and K-deficiency symptoms on corn, but heavy fertilization with these elements could not eliminate the symptoms. Allelopathy was suggested as a mechanism of interference in both of these studies.

B. Inhibition of Cell Division

A variety of allelochemics have been shown to inhibit mitosis in plant roots.[9] For example, coumarin can completely block mitosis in onion (*Allium cepa* L.) roots within a few hours after treatment.[123] Volatile terpenes from *Salvia leucophylla* are potent inhibitors of mitosis in cucumber seedlings. These compounds are also inhibitory to division of a large number of bacterial species isolated from soil.[124] Research conducted in the U.S.S.R. indicates that toxins from perennial sowthistle (*Sonchus arvensis* L.), common lambsquarters, and Canada thistle can reduce mitotic activity in roots of several crops.[125]

C. Inhibition of Extension Growth

The plant growth hormones indoleacetic acid (IAA) and the gibberellins (GA) have been shown to regulate cell enlargement in plants. IAA is present in plants in both active and inactive forms and is inactivated by IAA oxidase. Several allelopathic compounds have been shown to influence the activation of IAA oxidase. For example, monohydroxybenzoic acids are stimulatory to IAA oxidase, whereas 3,4-dihydroxybenzoic acid and ferulic acid are strongly inhibitory.[126] A number of investigations support the idea that both the phenolic compounds and flavonoids glycosides play a key role in the regulation of IAA oxidase activity in plants.[9] Other inhibitors have been reported to block GA-induced extension growth. Among these are phenolic compounds and hydrolyzable tannins.[9]

D. Inhibition of Photosynthesis

Many of the commercially marketed synthetic herbicides block this vital process extremely early in the life of the weed seedling. In fact, the seedlings are killed as soon as they perceive light, which often occurs just before they emerge. Photosynthetic inhibitors may be electron-transport inhibitors, uncouplers, energy-transfer inhibitors, electron acceptors, or may function by various combinations of the above.[127] There are only limited reports concerning the influence of allelopathic agents on the photosynthetic reactions. One study demonstrated a reduction in photosynthesis in sunflower, tobacco, and pigweed soon after treatment with scopoletin, a coumarin compound.[128] This compound also causes stomates to close, but it is not clear which process is affected first. Two recent studies indicated that several of the phenolic acids may reduce the chlorophyll content and photosynthetic rate in soybean plants.[129,130] Relatively high concentrations (5×10^{-4} to 10^{-3} *M*) were required in both studies. Although numerous synthetic quinones have been shown to inhibit photosynthesis, there are, to date, no reported effects of naturally occurring quinones on this process.[9] Phlorizin, a flavonoid from apple roots, is a potent energy-transfer inhibitor.[131]

E. Effects on Respiration

Chemicals may stimulate or inhibit respiration, both of which may be harmful to this energy-producing process. In the case of stimulation (enhanced O_2 uptake) the oxidative phosphorylation sequence may be uncoupled, resulting in a lack of ATP (energy) formation. Juglone has been shown to uncouple oxidative phosphorylation, as have a variety of aromatic acids, phenolics, aldehydes, flavonoids, and coumarin compounds.[9] Cyanogenesis may provide levels of cyanide that can uncouple oxidative phosphorylation in sensitive plant roots.

Several compounds isolated from soils have been shown to inhibit the respiration of plant roots. Juglone is particularly effective in this regard, causing more than a 90% reduction in the respiration of corn roots after 1-h exposure.[132] In addition, volatile terpenes from *Salvia* species have been shown to markedly reduce respiration in mitochondria isolated from oats and cucumbers.[133]

F. Effects on Protein Synthesis

Studies which monitor the influence of allelochemics on protein synthesis have utilized radiolabeled (^{14}C) sugars or amino acids and traced incorporation of the label into protein. Ferulic acid and coumarin were both found to inhibit the incorporation of ^{14}C-phenylalanine into the protein of seeds and embryos.[134] Similarly, incorporation of ^{14}C (from labeled CO_2) into protein was inhibited in algae by quinones.[135] One of the more important groups of synthetic herbicides, the acylanilides, also exert their action on protein synthesis in susceptible species.

G. Changes in Membrane Permeability

Many biologically active compounds are believed to exert their action through an alteration in the permeability characteristics of membranes. Plant membranes have been extremely difficult to study, although exudation of compounds from roots on root slices have been utilized as an index of permeability. The data of Balke and co-workers suggests that the phenolic compounds enhance efflux of K^+ from root tissues.[119] They propose that the initial site of action of salicylic acid is the plasmalemma, and that at low pH, both the plasmalemma and tonoplast may become leaky, causing massive losses of K^+. Previously, several of the phenolic compounds had been shown to increase permeability of neuronal membranes (from mollusks) to ions, particularly K^+.[136] A number of plant pathogens that induce wilt diseases are believed to produce toxins, for example, fusaric acid or α-picolinic acid are thought to cause losses in membrane integrity.[110]

H. Inhibition of Enzyme Activity

A variety of plant enzymes are inhibited in the presence of allelochemics. For example, the phosphorylase of potatoes is inhibited by chlorogenic acid, caffeic acid, and catechol, and their concentrations in the peels are probably sufficient to completely inhibit the activity of the enzyme.[137] Tannins have been reported to inhibit the activity of peroxidase, catalase, cellulase, polygalacturonase, amylase, and a variety of other enzymes.[9]

Root secretions of crops have been reported to inhibit catalase and peroxidase activities in common lambsquarters and redroot pigweed.[138] One mechanism by which plants are believed to resist disease is to produce inhibitors that prevent action of various enzymes exuded by the causal organism. Inhibition of key enzymes necessary for seed germination could provide an effective herbicidal mode of action for natural products. Germination of weed seeds is apparently inhibited by a variety of secondary compounds.[7,36,97,99,139]

IX. IMPACT OF ALLELOCHEMICS IN NATURAL AND AGROECOSYSTEMS

A. Influence on Plant Succession

Plant succession, particularly in old fields and cut-over forests, has intrigued ecologists for many decades. The appearance and disappearance of species and changes in species dominance over time has been attributed to numerous factors including physical changes in the habitat, seed production and dispersal, competition for resources, or combinations of all these. Rice and co-workers[9] have presented extensive evidence that allelopathy may play an important role in the disappearance of the pioneer weeds (those most rapidly invading old fields). For example, johnsongrass and annual sunflowers revegetating old fields contain inhibitors that prevent germination of many other weed species, but have little or no effect on the germination of triple awn grass (*Aristida oligantha* Michx.), which often follows them in succession. Rice's group also found that large crabgrass (*Digitaria sanguinalis* [L.] Scop.) may inhibit the germination of several other species of annual weeds.

Certain reforestation problems have also been linked to allelopathy. There are logged-over sites on the Allegheny Plateau in northwestern Pennsylvania that have remained essentially treeless for up to 80 years.[25,26] Several herbaceous weed species have been shown to produce toxins that inhibit establishment of the black cherry (*Prunus serotina* Ehrh.) seedlings that normally reinfest these sites. Among the more active are goldenrods (*Solidago*) and *Aster* species.

B. Inhibition of Nitrogen Fixation and Nitrification

Rice and associates[141] have contributed greatly in this area of research. Their hypothesis that succession proceeds toward selection of plants that inhibit nitrification and conserve ammonium nitrogen for their use has been well substantiated. They demonstrated extremely low populations of *Nitrosomonas* and *Nitrobacter*, two bacterial species responsible for nitrification reactions in soil areas where climax vegetation is present.

Rice also assayed numerous weed species from old fields for their ability to inhibit the nitrogen-fixing bacteria, *Azotobacter* and *Rhizobium*.[9] The latter group of bacteria are particularly important to legume crop plants and ultimately to mankind, because they infect them, produce nodules, and then proceed to fix atmospheric nitrogen, which is utilized for crop growth. Extracts of several important weed species were found to inhibit the nodulation of legumes by *Rhizobium*. Among those were western ragweed, large crabgrass, prostrate spurge, and annual sunflower. More recent studies also indicate that *Aristida adscensionis* produces compounds that are inhibitory to N-fixing organisms.[142] Adverse effects of weeds on N-fixation may be an agricultural problem that deserves much more research attention.

C. Patterning of Plant Species

In many ecosystems, plants tend to pattern themselves as pure stands or as individuals spaced in rather specific densities or configurations. Many desert species show obvious zones of inhibition around which few, if any, aliens are allowed to invade.[1] These patterns often cannot be adequately explained by competition alone, and are probably caused by a combination of factors including allelopathy. The phenomenon happens with herbaceous plants as well as woody shrubs and trees. In addition to the desert and chaparral species previously mentioned, black walnut, black locust (*Robinia pseudoacacia* L.), red pine (*Pinus densi-*

flora), eucalyptus (several species), and sycamore (*Platanus occidentialis* L.) seem to exert selective herbicidal influences on herbaceous vegetation in their immediate vicinity.[9]

Muller reported that the black mustard (*Brassica nigra* [L.] Koch.) can form almost pure stands after invading annual grasslands of coastal southern California.[1] This was attributed to inhibitors released from the dead stalks and leaves that do not allow germination and growth of other plants. These observations provide agronomists hope that similar results could be exploited with crops, specifically to achieve almost pure stands of crops (over weeds) by use of an allelopathic mechanism.[14]

D. Prevention of Seed Decay and Germination

Weed seed can lie dormant in the soil for many years but still remain viable. The classic seed burial study of Beal and successors have shown seeds of at least one weed species, Moth Mullein (*Verbascum blattaria* L.), can remain viable in soil for a period of 100 years.[143] Three species continued to germinate after 80 years of burial. Weed seeds not only resist decay by soil microbes, but they vary in dormancy characteristics. There is considerable evidence that chemical inhibitors are responsible for both phenomena. Unsaturated lactones and phenolic compounds, in particular, are potent antimicrobial compounds and are present in many seeds.[9] Fruits and seeds are also known to contain diverse germination inhibitors including phenolic compounds, flavonoids and/or their glycosides, and tannins.

E. Replant Problems

Several plants are known to release inhibitors into soil that cause problems with successful replanting of crops into that area. Perhaps the most famous of these are the orchard replant problems, particularly peach[24] and apple.[23] In both of these cases, there appear to be several reasons for the toxicity exerted on replanted trees, but the release of toxins by the decaying root residues has been clearly demonstrated.

At the beginning of this century, several researchers noted that croplands upon which sorghum was grown repeatedly soon became unproductive.[21,22] Even at that early date, part of the problem was attributed to chemical influences. Since that time, toxins have been identified from a variety of cereal grain residues.[10]

X. EXPLOITATION OF ALLELOPATHY IN WEED MANAGEMENT

It has been previously demonstrated by ecologists that allelopathy may influence plant distribution patterns. Recently, some weed scientists have attempted to exploit this idea as an additional weed management strategy. One approach has been to screen for allelopathic types in germplasm collections of crops, the idea being to transfer this character into cultivars by either conventional breeding or other genetic-transfer techniques. Superior weed-suppressing types have been reported from searches of cucumber,[113] oat,[105] sunflower,[140] and soybean collections.[144,145] In the case of cucumber, toxic leachates were produced that could be neutralized with activated charcoal.[7] Inconsistent results were obtained on fine-textured soils or after heavy rainfall.[146] When thoroughly researched, this idea could hold promise for crop plants that are maintained in high-density monocultures, i.e., turfgrasses, forage grasses, or legumes.

Another approach has been to utilize allelopathic rotational crops or companion plants in annual or perennial cropping systems.[48,49] Living rye (*Secale cereale* L.) and its residues have been shown to provide nearly complete suppression of a variety of agroecosystem weeds.[49] Similarly, residues of sorghums, barley, wheat, and oats can provide exceptional suppression of certain weed species.[48,147,148] Crops vary in their responses to allelopathic residues. Large-seeded legumes have often responded favorably, whereas small-seeded vegetables have been severely injured.[147] The use of cover crops and their residues may complement the adoption of conservation tillage practices in crop production. The plant residues do provide water and soil conservation as well as suppression.

Allelopathic plants may also provide a strategy for vegetation management in aquatic systems. The diminutive spikerush (*Eleocharis coloradoensis*) has been reported to displace more vigorous and unwanted aquatic plants, i.e., pondweeds (*Potamogenton* species) and *Elodea* in canals and drainage ditches.[149,150] Frank[151] attributed this to allelopathic effects, and more recently the phytotoxic compound dihydroactinidiolide (DAD) was isolated from the plant.[152] This chemical has since been shown to be inhibitory to pondweeds.

It is likely that allelopathy may account for a major portion of interference by some species of weeds and have little or no impact with others. We hypothesize that allelopathy may be implicated in shifts in species distribution and increases of dominance by certain annual and perennial weeds in a variety of crop ecosystems. Research efforts should be directed toward identification of weed species that are allelopathic and growth stages when toxin production occurs.

NOTE ADDED IN PROOF

Since the original preparation of the manuscript, the author has become aware of more recent reports of compounds with allelopathic potential. These include quinolizidine alkaloids from *Lupinus* species,[153] carboxyatractyloside from *Xanthium strumarium* L.,[154] asperuloside from *Galium aparine* L.,[155] and taxiphyllin, an epimer of dhurrin, from Johnsongrass. In addition, evidence has been presented for allelopathy in diffuse knapweed (*Centaurea diffusa* Lam),[156] and vinegar weed (*Trichostema lancealotum*).[157]

REFERENCES

1. **Muller, C. H.,** Allelopathy as a factor in ecological process, *Vegetatio*, 18, 348, 1969.
2. **Bykov, B. A.,** Vedenie v. fitocenologiju. "Nauka" Alma-Ata, *Introduction to Phytocoenology* (in Russian), Kiev, 1970.
3. **Molisch, H.,** *Der Einfluss einer pflanze auf die andere — Allelopathie*, G. Fischer, Jena, 1937.
4. **Szczepanski, A. J.,** Allelopathy as a means of biological control of water weeds, *Aquatic Bot.*, 3, 193, 1977.
5. **Harper, J. L.,** *Population Biology of Plants*, Academic Press, New York, 1977.
6. **Tang, C. H. and Young, C.-C.,** Collection and identification of allelopathic compounds from the undisturbed root system of Bigalta limpograss *(Hemarthria altissima), Plant Physiol.*, 69, 155, 1982.
7. **Lockerman, R. H. and Putnam, A. R.,** Growth inhibitors in cucumber plants and seeds, *J. Am. Soc. Hortic. Sci.*, 106, 418, 1981.
8. **Muller, C. H., Muller, W. H., and Haines, B. L.,** Volatile growth inhibitors produced by shrubs, *Science*, 143, 471, 1964.
9. **Rice, E. L.,** *Allelopathy*, Academic Press, New York, 1974, 1.
10. **Smith, A. M. and Secoy, D. M.,** Pest controls of the ancients, *Furrow*, p. 27, 1977.
11. **deCandolle, M. A. P.,** *Physiologie Vegetale*. III. Bechet Jeune, Lib. Fac. Med. 1474, Paris, 1832.
12. **McCalla, T. M. and Haskins, F. A.,** Phytotoxic substances from soil microorganisms and crop residues, *Bacterial. Rev.*, 28, 181, 1964.
13. **Patrick, Z. A.,** Phytotoxic substances associated with the decomposition in soil of plant residues, *Soil Sci.*, 111, 13, 1971.
14. **Putnam, A. R. and Duke, W. B.,** Allelopathy in agroecosystems, *Annu. Rev. Phytopathol.*, 16, 431, 1978.
15. **Altieri, M. A. and Doll, J. D.,** The potential of allelopathy as a tool for weed management in crop fields, *PANS (Pest Artic. News Summ.)*, 24, 495, 1978.
16. **Bell, D. T. and Koeppe, D. E.,** Non-competitive effects of giant foxtail on the growth of corn, *Agron. J.*, 64, 321, 1972.
17. **Colton, C. E. and Einhellig, F. A.,** Allelopathic mechanisms of velvetleaf (*Abutilon theophrasti* medic., *Malvaceae*) on soybean, *Am. J. Bot.*, 67, 407, 1980.

18. **Kommedahl, T., Kotheimer, J. B., and Bernardini, J. V.,** Effects of quackgrass on germination and seedling development of certain crop plants, *Weeds*, 7, 1, 1959.
19. **Rasmussen, J. A. and Einhellig, F. A.,** Noncompetitive effects of common milkweed (*Asclepias syriaca* L.) on germination and growth of grain sorghum, *Am. Midl. Nat.*, 94, 478, 1975.
20. **Guenzi, W. D. and McCalla, T. M.,** Inhibition of germination and seedling development by crop residues, *Soil Sci. Soc. Am. Proc.*, 26, 456, 1962.
21. **Conrad, J. P.,** Some causes of the injurious after effects of sorghum and suggested remedies, *J. Am. Soc. Agron.*, 19, 1091, 1927.
22. **Breazeale, J. B.,** The injurious after effects of sorghum, *Agron. J.*, 16, 689, 1924.
23. **Borner, H.,** The apple replant problem. I. The excretion of phlorizen from apple root residues, *Contrib. Boyce Thompson Inst.*, 20, 39, 1959.
24. **Proebsting, E. L.,** A case history of a "peach replant" situation, *Proc. Am. Soc. Hortic. Sci.*, 38, 21, 1950.
25. **Horsley, S. B.,** Allelopathic inhibition of black cherry by fern, grass, goldenrod, and aster, *Can. J. For. Res.*, 7, 205, 1977.
26. **Horsley, S. B.,** Allelopathic inhibition of black cherry. II. Inhibition by woodland grass, ferns, and club moss, *Can. J. For. Res.*, 7, 515, 1977.
27. **Bode, H. R.,** Uber die Blattausscheidungen des wermuts and ihre Werkung auf andere Pflanzen, *Planta*, 30, 567, 1940.
28. **Bell, D. T. and Muller, C. H.,** Dominance of California annual grasslands by *Brassica nigra*, *Am. Midl. Nat.*, 90, 277, 1973.
29. **Tinnin, R. and Muller, C. H.,** The allelopathic potential of *avena fatua:* influence on herb distribution, *Bull. Torrey Bot. Club*, 98, 243, 1971.
30. **Lee, I. K. and Monsi, M.,** Ecological studies on *Pinus densiflora* forest. I. Effects of plant substances on the floristic composition of the undergrowth, *Bot. Mag.*, 76, 400, 1963.
31. **delMoral, R. and Muller, C. H.,** The allelopathic effects of *Eucalyptus canaldulensis*, *Am. Midl. Nat.*, 83, 254, 1970.
32. **Fuerst, E. P. and Putnam, A. R.,** Separating the competitive and allelopathic components of interference: theoretical principles, *J. Chem. Ecol.*, 9, 1045, 1983.
33. **Whittaker, R. H. and Feeny, P. P.,** Allelochemics: chemical interaction between species, *Science*, 171, 757, 1971.
34. **Grummer, G.,** *Die gegenseitige Beeinflussung hoherer Pflanzen-Allelopathie*, G. Fischer, Jena, 1955.
35. **Stowe, L. G.,** Allelopathy and its influences on the distribution of plants in an Illinois old-field, *J. Ecol.*, 67, 1065, 1979.
36. **Evanari, M.,** Germination inhibitors, *Bot. Rev.*, 15, 153, 1949.
37. **Rasmussen, J. A. and Einhellig, F. A.,** Inhibitory effects of combinations of three phenolic acids on grain sorghum germination, *Plant Sci. Lett.*, 14, 69, 1979.
38. **Muller, C. H.,** Inhibitory terpenes volatilized from *Salvia* shrubs, *Bull. Torrey Bot. Club*, 92, 38, 1965.
39. **Woods, F. W.,** Biological antagonisms due to phytotoxic root exudates, *Bot. Rev.*, 26, 546, 1960.
40. **Rovira, A. D.,** Plant root exudates, *Bot. Rev.*, 35, 35, 1969.
41. **Caussanel, J.-P. and Kunesch, G.,** Qualitative and quantitative study of growth inhibitors in root exudates of common lambsquarters (*Chenopodium album* L.) at the beginning of its flowering on hydroponic culture and under controlled conditions, *Z. Pflanzenphysiol.*, 93, 229, 1979.
42. **McDougall, B. M. and Rovira, A. D.,** Carbon-14 labeled photosynthate in wheat root exudates, *Nature (London)*, 207, 1104, 1965.
43. **Preston, W. H., Mitchell, J. W., and Rieve, W.,** Movement of alpha-methoxy-phenyl- acetic acid from one plant to another through the root systems, *Science*, 119, 437, 1954.
44. **Tukey, H. B., Jr.,** Leaching of metabolites from above ground plant parts and its implications, *Bull. Torrey Bot. Club*, 93, 385, 1966.
45. **Grummer, G. and Beyer, H.,** The influence exerted by species of *Camelina* on flax by means of toxic substances, in *The Biology of Weeds*, Harper, J. L., Ed., Blackwell Scientific, Oxford, 1960, 153.
46. **Abdul-Wahab, A. S. and Rice, E. L.,** Plant inhibition by Johnson grass and its possible significance in old-field succession, *Bull. Torrey Bot. Club*, 94, 486, 1967.
47. **Proebsting, E. L. and Gilmore, A. E.,** The relation of peach root toxicity to the reestablishing of peach orchards, *Proc. Am. Soc. Hortic. Sci.*, 38, 21, 1941.
48. **Putnam, A. R. and DeFrank, J.,** Use of phytotoxic plant residues for selective weed control, *Crop Prot.*, 2, 173, 1983.
49. **Barnes, J. P. and Putnam, A. R.,** Rye residues contribute weed suppression in no-tillage cropping systems, *J. Chem. Ecol.*, 9, 1045, 1983.
50. **Gabor, W. E. and Veatch, C.,** Isolation of a phytotoxin from quackgrass *Agropyron repens* rhizomes, *Weed Sci.*, 29, 155, 1981.

51. **Wilson, R. G., Jr.**, Effect of Canada thistle (*Cirsium arvense*) residues on growth of some crops, *Weed Sci.*, 29, 159, 1981.
52. **Kovacs, M. F., Jr.**, Dhurrin (*P*-hydroxy-mandelonitrile-*B*-D-glucoside): an Allelopath Identified in Johnsongrass (*Sorghum halepense* (L.) pers.) Rhizome Exudate, Ph.D. dissertation, University of Maryland, College Park, 1972.
53. **Drost, D. C. and Doll, J. D.**, The allelopathic effect of yellow nutsedge on corn and soybeans, *Weed Sci.*, 28, 229, 1980.
54. **Schreiber, M. M. and Williams, J. L.**, Toxicity of root residues of weed grass species, *Weeds*, 15, 80, 1967.
55. **Gressel, J. B. and Holm, L. G.**, Chemical inhibition of crop germination by weed seeds and the nature of inhibition by *Abutilon theophrasti*, *Weed Res.*, 4, 44, 1964.
56. **Welbank, P. J.**, Toxin production during decay of *Agropyron repens* and other species, *Weed Res.*, 3, 205, 1963.
57. **Gajic, D. and Nikocevic, G.**, Chemical allelopathic effect of *Agrostemma githago* upon wheat, *Fragm. Herb. Jugosl.*, 18, 1, 1973.
58. **Osvald, H.**, An antagonism between plants, in *Intern. Congr. Bot. (7th) Proc.*, Stockholm, 1950.
59. **Van der Veen, R.**, Root competition in coffee and rubber plantations, *Arch. Koffiecult (Nederland India)*, 3, 65, 1935.
60. **Jackson, J. R. and Willemsen, R. W.**, Allelopathy in the first stages of secondary succession on the piedmont of New Jersey, *Am. J. Bot.*, 7, 1015, 1976.
61. **Steenhagen, D. A. and Zimdahl, R. L.**, Allelopathy of leafy spurge *(Euphorbia esula)*, *Weed Sci.*, 27, 1, 1979.
62. **Letourneau, D., Failes, G. D., and Heggeness, H. G.**, The effect of aqueous extracts of plant tissue on germination of seeds and growth of seedlings, *Weeds*, 4, 363, 1956.
63. **Funke, G. L.**, The influence of *Artemisia absinthium* on neighboring plants, *Blumea*, 5, 281, 1943.
64. **Eussen, J. H. H.**, Isolation of growth inhibiting substances from alang-alang (*Imperata cylindrical* (L.) Beauv., in *Studies on the Tropical Weed Imperata*, Papers No. 6 and 7, Drukkerig Elinkwijk BV, Utrecht, 1978.
65. **Pandya, S. M.**, Effect of *Celosia argentea* extracts on root and shoot growth of bjra seedlings, *Geobios (Jodhpur)*, 2, 175, 1975.
66. **Stachon, W. J. and Zimdahl, R. L.**, Allelophatic activity of Canada thistle (*Cirsium arvense*) in Colorado, *Weed Sci.*, 28, 83, 1980.
67. **Tames, R. S., Gesto, M. D. V., and Vieitez, E.**, Growth substances isolated from tubers of *Cyperus esculentus* var. *aureus*, *Physiol. Plant.*, 28, 195, 1973.
68. **Lucena, J. M. and Doll, J. D.**, Efectos de inhibidores de crecimiento del coquito (*Cyperus rotundus*) sobre sorgoy soya, *Rev. Comalfi*, 3, 257, 1976.
69. **Friedman, T. and Horowitz, M.**, Biologically active substances in subterranean parts of purple nutsedge, *Weed Sci.*, 19, 398, 1971.
70. **Sarma, K. K. V.**, Allelopathic potential of *Digera arvensis* Forsk. on *Pennisetum typhoides* Stapf. et Hubb, *Geobios (Jodhpur)*, 1, 137, 1974.
71. **Neill, R. L. and Rice, E. L.**, Possible role of *Ambrosia psilostachya* on patterning and succession in old-fields, *Am. Midl. Nat.*, 86, 344, 1971.
72. **Parenti, R. L. and Rice, E. L.**, Inhibitional effects of *Digitaria sanguinalis* and possible role in old-field succession, *Bull. Torrey Bot. Club*, 96, 70, 1969.
73. **Ballester, A., Albo, H. M., and Vieitez, E.**, The allelopathic potential of *Erica scoparia* L., *Oecologia*, 30, 55, 1977.
74. **Letourneau, D. and Heggeness, H. G.**, Germination and growth inhibitors in leafy spurge foliage and quackgrass rhizomes, *Weeds*, 4, 12, 1957.
75. **Brown, D. D.**, The Possible Ecological Significance of Inhibition by *Euphorbia supina*, M.S. thesis, University of Oklahoma, Norman, 1968.
76. **Anderson, R. C., Katz, A. J., and Anderson, M. R.**, Allelopathy as a factor in the success of *Helianthus mollis; Lam, J. Chem. Ecol.*, 4, 9, 1978.
77. **Mann, H. H. and Barnes, T. W.**, The competition between barley and certain weeds under controlled conditions. II. Competition with *Holcus mollis*, *Ann. Appl. Biol.*, 34, 252, 1947.
78. **Naqvi, H. H. and Muller, C. H.**, Biochemical inhibition (allelopathy) exhibited by Italian ryegrass (*Lolium multiflorum* L.), *Pak. J. Bot.*, 7, 139, 1975.
79. **Mann, H. H. and Barnes, T. W.**, The competition between barley and certain weeds under controlled conditions, *Ann. Appl. Biol.*, 32, 15, 1945.
80. **Sarma, K. K. V.**, Allelopathic potential of *Parthenium hysterophorus* Linn. on seed germination, and dry matter production in *Arachis hypogea* Willd., *Crotoloria juncea* Linn. and *Phaseolus mungo* Linn., *Trop. Ecol.*, 17, 76, 1976.

81. **Kanchan, S. D.,** Allelopathic effects of *Parthenium hysterophorus* L. (tropical American weed). I. Exudation of inhibitors through roots, *Plant Soil,* 53, 27, 1979.
82. **Alderman, W. H. and Middleton, J. A.,** Toxic relations of other crops to tomatoes, *Proc. Am. Soc. Hortic. Sci.,* 22, 307, 1925.
83. **Datta, S. C. and Chatterjee, A. K.,** Allelopathy in *Polygonum orientale:* inhibition of seed germination and seedling of mustard, *Comp. Physiol. Ecol.,* 5, 54, 1980.
84. **Martin, P. and Rademacher, B.,** Studies on the mutual influences of weeds and crops, in *Biology of Weeds,* Harper, J. L., Ed., Blackwell Scientific, Oxford, 1960, 143.
85. **Einhellig, F. A. and Rasmussen, J. A.,** Allelopathic effects of *Rumex crispus* on *Amaranthus retroflexus,* grain sorghum, and field corn, *Am. Midl. Nat.,* 90, 79, 1973.
86. **Amritphale, D. and Mall, L. P.,** Allelopathic influence of *Saccharum spontaneum* L. on the growth of three varieties of wheat, *Sci. Cult.,* 44, 28, 1978.
87. **Lodhi, M. A. K.,** Allelopathic potential of *Salsola kali* L. and its possible role in rapid disappearance of weedy stage during revegetation, *J. Chem. Ecol.,* 5, 429, 1979.
88. **Anaya, A. L. and Gomez-Pompa, A.,** Inhibicion del crecimiento prodecida por el "piru" (*Schinus molle* L.), *Rev. Soc. Mex. Hist. Nat.,* 32, 99, 1971.
89. **Mann, H. H. and Barnes, T. W.,** The competition between barley and certain weeds under controlled conditions. IV. Competition with *Stellaria media, Ann. Appl. Biol.,* 37, 139, 1950.
90. **Rice, E. L.,** Allelopathy — an update, *Bot. Rev.,* 45, 15, 1979.
91. **Horsley, S. B.,** Allelopathic interference among plants. II. Physiological modes of action, in *Proc. 4th North American Forest Biology Workshop,* Wilcox, H. E. and Hamer, A. F., Eds., College of Environmental Science and Forestry, Syracuse, N.Y., 1977, 93.
92. **Conn, E. E.,** Cyanogenic compounds, *Annu. Rev. Plant Physiol.,* 31, 433, 1980.
93. **Barton, L. V.,** Toxicity of ammonia, chlorine, hydrogen cyanide, hydrogen sulfide, and sulfur dioxide gases. IV. Seeds, *Contr. Boyce Thompson Inst.,* 11, 351, 1939.
94. **Stout, M. and Tolman, B.,** Factors affecting the germination of sugarbeet and other seeds with special reference to the toxic effects of ammonia, *J. Agric. Res.,* 63, 687, 1941.
95. **Hansen, E.,** Quantitative study of ethylene production in apple varieties, *Plant Physiol.,* 20, 631, 1945.
96. **Eplee, R. E.,** Ethylene: a witchweed seed germination stimulant, *Weed Sci.,* 23, 433, 1975.
97. **Lehle, F. R. and Putnam, A. R.,** Allelopathic potential of sorghum (*Sorghum bicolor*): isolation of seed germination inhibitors, *J. Chem. Ecol.,* 9, 1223, 1983.
98. **Lynch, J. M., Hall, K. C., Anderson, H. A., and Hepburn, A.,** Organic acids from the anaerabic decomposition of *Agropyron repens* rhizomes, *Phytochemistry,* 19, 1846, 1980.
99. **Lynch, J. M.,** Effects of organic acids on the germination of seeds and growth of seedlings, *Plant Cell Environ.,* 3, 255, 1980.
100. **Mojc, W.,** Organic soil toxins, in *Diagnostic Criteria for Plants and Soils,* Chapman, H. D., Ed., University of California, Riverside, 1966, chap. 37.
101. **Elmore, C. D.,** Free amino acids of *Abutilon theophrasti* seed, *Weed Res.,* 20, 63, 1980.
102. **Norstadt, F. A. and McCalla, T. M.,** Phytotoxic substance from a species of *Penicillium, Science,* 140, 410, 1963.
103. **Winter, A. G.,** New physiological and biological aspects in the interrelationship between higher plants, *Symp. Soc. Exp. Biol.,* 15, 229, 1961.
104. **Avers, C. J. and Goodwin, R. H.,** Studies on roots. IV. Effects of coumarin and scopoletin on the standard root growth pattern of *Phleum pratense, Am. J. Bot.,* 43, 612, 1956.
105. **Fay, P. K. and Duke, W. B.,** An assessment of allelopathic potential in *Avena* germplasm, *Weed Sci.,* 25, 224, 1977.
106. **Davis, R. F.,** The toxic principle of *Juglans nigra* as identified with synthetic juglone and its toxic effects on tomato and alfalfa plants, *Am. J. Bot.,* 15, 620, 1928.
107. **Kohlmuenzer, S.,** Botanical and chemical studies of the collective species *Galium mollugo* with reference to karyotypes growing in Poland. V. Phytochemical studies, *Diss. Pharm.,* 17, 357, 1965.
108. **Rice, E. L. and Pancholy, S. K.,** Inhibition of nitrification by climax ecosystems. II. Additional evidence and possible role of tannins, *Am. J. Bot.,* 60, 691, 1973.
109. **Rizvi, S. J., Mukerji, H. D., and Mathur, S. N.,** Selective phytotoxicity of 1,3,7-trimethyl xanthine between *Phaseolus mungo* and some weeds, *Agric. Biol. Chem.,* 45, 1255, 1981.
110. **Owens, L. D.,** Toxins in plant disease: structure and mode of action, *Science,* 165, 18, 1969.
111. **Swain, T.,** Secondary compounds as protective agents, *Annu. Rev. Plant Physiol.,* 28, 479, 1977.
112. **Melrod, D. M.,** Allelopathy: a Methodological Approach into Root Exudations, M.S. thesis, Cornell University, Ithaca, N.Y., 1977.
113. **Putnam, A. R. and Duke, W. B.,** Biological suppression of weeds: evidence for allelopathy in accessions of cucumber, *Science,* 185, 370, 1974.
114. **Frey-Wyssling, A. and Babler, S.,** Zur biochemie des Gewachshaustabaks, *Experientia,* 13, 399, 1957.

115. **Koeppe, D. E., Rohrbaugh, L. M., and Wender, S. H.,** The effect of varying UV intensities on the concentrations of scopolin and caffeoylquinic acids in tobacco and sunflower, *Phytochemistry*, 8, 889, 1969.
116. **Tso, T. C., Kasperbauer, N. J., and Sorokin, T. P.,** Effects of photoperiod and end-of-day light quality on alkaloids and phenolic compounds of tobacco, *Plant Physiol.*, 45, 330, 1970.
117. **del Moral, R.,** On the variability of chlorogenic acid concentration, *Oecologia*, 9, 289, 1972.
118. **Dieterman, L. J., Lin, C.-Y., Rohrbaugh, L. M., and Wender, S. H.,** Accumulation of ayapin and scopolin in sunflower plants treated with 2,4-dichlorophenoxyacetic acid, *Arch. Biochem. Biophys.*, 106, 275, 1964.
119. **Harper, J. R. and Balke, N. E.,** Characterization of the inhibition of K^+ absorption in oat roots by salicylic acid, *Plant Physiol.*, 68, 1349, 1981.
120. **Harper, J. R. and Balke, N. E.,** Inhibition of potassium absorption in excised oat roots by phenolic acids, in *Plant Membrane Transport: Current Conceptual Issues*, Spanswick, R. M., Lucas, W. J., and Dainty, J., Eds., Elsevier/North Holland, New York, 1980, 399.
121. **Chambers, E. E. and Holm, L. G.,** Phosphorus uptake as influenced by associated plants, *Weeds*, 13, 312, 1965.
122. **Buchholtz, K. P.,** The influence of allelopathy on mineral nutrition, in *Biochemical Interactions Among Plants*, National Academy of Sciences, Washington, D.C., 1971, 86.
123. **Cornman, I.,** Alteration of mitosis by coumarin and parasorbic acid, *Am. J. Bot.*, 33, 217, 1946.
124. **Muller, W. H.,** Volatile materials produced by *Salvia leucophylla:* effects on seedling growth and soil bacteria, *Bot. Gaz.*, 126, 195, 1965.
125. **Bukolova, T. P.,** A study of the mechanism of action of water-soluble substances of weeds on cultivated plants, in *Physiological-Biochemical Basis of Plant Interaction in Phytocenoses*, Grodzinsky, N. M., Ed., Naukova Dumka, Kiev, 2, 1971, 66.
126. **Lee, T. T. and Skoog, F.,** Effects of hydroxybenzoic acids on indoleacetic acid inactivation by tobacco callus extracts, *Physiol. Plant.*, 18, 577, 1965.
127. **Moreland, D. E.,** Mechanisms of action of herbicides, *Annu. Rev. Plant Physiol.*, 31, 597, 1980.
128. **Einhellig, F. A., Rice, E. L., Risser, P. C., and Wender, S. H.,** Effects of scopoletin on growth, CO_2 exchange rates, and concentrations of scopoletin, scopolin, and chlorogenic acid in tobacco, sunflower and pigweed, *Bull. Torrey Bot. Club*, 97, 22, 1970.
129. **Einhellig, F. A. and Rasmussen, J. A.,** Effects of three phenolic acids on chlorophyll content and growth of soybean and grain sorghum seedlings, *J. Chem. Ecol.*, 5, 815, 1979.
130. **Patterson, D. T.,** Effects of allelopathic chemicals on growth and physiological responses of soybeans *(Glycine max.)*, *Weed Sci.*, 29, 53, 1981.
131. **McCarty, R. E.,** Delineation of the mechanism of ATP synthesis in chloroplasts: use of uncouplers, energy transfer inhibitors, and modifiers of coupling factor, *Methods Enzymol.*, 69, 723, 1980.
132. **Koeppe, D. E.,** Some reactions of isolated corn mitochondria influenced by Juglone, *Physiol. Plant*, 27, 89, 1972.
133. **Muller, W. H., Lorber, D., Haley, B., and Johnson, K.,** Volatile growth inhibitors produced by *Salvia leucophylla:* effect of oxygen uptake by mitochondrial suppressions, *Bull. Torrey Bot. Club*, 96, 89, 1969.
134. **Van Sumere, C. F., Cottenie, J., DeGreef, J., and Kint, J.,** Biochemical studies in relation to the possible germination regulatory role of naturally occurring coumarin and phenolics, *Recent Adv. Phytochem.*, 4, 165, 1971.
135. **Zweig, G., Carroll, J., Tamas, I., and Sikka, H. C.,** Studies on effects of certain quinones. II. Photosynthetic incorporation of $^{14}CO_2$ by *Chlorella*, *Plant Physiol.*, 49, 385, 1972.
136. **Levitan, H. and Barker, J. L.,** Salicylate: a structure-activity study of its effects on membrane permeability, *Science*, 176, 1423, 1972.
137. **Schwimmer, S.,** Influence of polyphenols and potato components on potato phosphorylase, *J. Biol. Chem.*, 232, 715, 1958.
138. **Dzubenko, N. N. and Petrenko, N. I.,** On biochemical interaction of cultivated plants and weeds, in *Biochemical and Physiological Bases for Plant Interactions in Phytocenosis*, Grodzinsky, A. M., Ed., Naukova Dumka, Kiev, 2, 1971, 60.
139. **Williams, R. D. and Hoagland, R. E.,** The effects of naturally occurring phenolic compounds on seed germination, *Weed Sci.*, 30, 206, 1982.
140. **Leather, G. R.,** Sunflowers (*Helianthus annus*) are allelopathic to weeds, *Weed Sci.*, 31, 37, 1983.
141. **Rice, E. L. and Pancholy, S. K.,** Inhibition of nitrification by climax ecosystems, *Am. J. Bot.*, 59, 1033, 1972.
142. **Murthy, M. S. and Nagodra, T.,** Allelopathic effects of *Aristida adscensionis* on *Rhizobium*, *J. Appl. Ecol.*, 14, 279, 1977.
143. **Kivilaan, A. and Bandurski, R. S.,** The ninety-year period for Dr. Beal's seed viability experiment, *Am. J. Bot.*, 60, 146, 1973.

144. **Massantini, F., Caporali, F., and Zellini, G.,** Evidence for allelopathic control of weeds in lines of soybeans, *Symp. Different Methods Weed Control Integration,* 1, 23, 1977.
145. **Rose, S. J., Burnside, O. C., Specht, J. E., Williams, J. H., and Swisher, B. A.,** Competition and allelopathy of soybean (Glycine max (L.) Merr.) cultivars to weeds, *Weed Sci. Soc. Am.,* Abstr., p. 16, 1983.
146. **Lockerman, R. H. and Putnam, A. R.,** Field evaluation of allelopathic cucumbers as an aid to weed control, *Weed Sci.,* 27, 54, 1978.
147. **Putnam, A. R., DeFrank, J., and Barnes, J. P.,** Exploitation of allelopathy for weed control in annual and perennial cropping systems, *J. Chem. Ecol.,* 9, 1001, 1983.
148. **Steinsiek, T. W., Oliver, L. R., and Collins, F. C.,** Allelopathic potential of wheat (*Triticum aestivum*) straw on selected weed species, *Weed Sci.,* 30, 495, 1982.
149. **Osborn, E. T., Moran, W. T., Greene, K. T., and Bartley, T. R.,** Weed Control Investigations on Some Important Plants which Impede the Flow of Western Irrigation Waters, Joint Lab. Report, 51-2, Bureau of Reclamation, U.S. Department of Agriculture, ARS, Denver, Colo., 1954.
150. **Yeo, R. R. and Fisher, T. W.,** Progress and Potential for Biological Weed Control with Fish, Pathogens, Competitive Plants and Snails, Tech. Paper FAO Int. Conf. on Weed Control, Weed Science Society of America, Champaign, Ill., 688 p., 1970.
151. **Frank, P. A. and Dechoretz, N.,** Allelopathy in dwarf spikerush *(Eleocharis coloradoensis), Weed Sci.,* 28, 499, 1980.
152. **Stevens, K. L. and Merrill, G. B.,** Growth inhibitors for spikerush, *J. Agric. Food Chem.,* 78, 644, 1980.
153. **Wink, M.,** Inhibition of seed germination by quinolizidine alkaloids, *Planta,* 158, 365, 1983.
154. **Cutler, H. G. and Cole, R. J.,** Carboxyatractyloside: a compound from *Xanthium strumarium* and *Atractylis gummifera* with plant growth inhibiting properties. The probable "Inhibitor A", *J. Nat. Prod.,* 46, 609, 1983.
155. **Komai, K., Iwamura, J., and Ueki, K.,** Plant growth inhibitor in the seeds of catchweed, *Weed Res. (Japan),* 28, 205, 1983.
156. **Muir, A. D. and Majak, W.,** Allelopathic potential of diffuse knapweed (*Centaurea diffusa*) extracts, *Can. J. Plant Sci.,* 63, 989, 1983.
157. **Heisey, R. M. and Delwiche, C. C.,** A survey of California plants for water extractable and volatile inhibitors, *Bot. Gaz.,* 144, 382, 1983.

INDEX

A

D

H

I

J

K

L

Q

R

S

T

U

V

W

X

For Product Safety Concerns and Information please contact our EU representative GPSR@taylorandfrancis.com Taylor & Francis Verlag GmbH, Kaufingerstraße 24, 80331 München, Germany

Batch number: 10731074

Printed by Printforce, the Netherlands